Contents

1. Introduction
2. Article 233: The holographic Sun in the clouds
3. Article 234: Stellar Core in LASCO C3 image has a comet like characteristics
4. Article 235: Is there a Planet 9 in the outer reaches of our Solar System
5. Article 236: What are Stellar Cores?
6. Article 237: Hawaiian Island and mega-tsunamis
7. Article 238: Aliens exist
8. Article 239: Primary and secondary alien technology
9. Article 240: Planet X System effect on radioactive decay rate and heating of planets
10. Article 241: The Rapture at Pentecost
11. Article 242: Planet X System affecting earth van Allen Belt particle acceleration
12. Article 243: Earth hosting at least 3 Planet X System Objects
13. Article 244: The Planet X System destroyer of Star Systems
14. Article 246: The ravaging of the nation of South Africa
15. Article 247: The effect of the Planet X System on the earth
16. Article 248: Planet X System Objects Interacting with earth
17. Article 249: Gravity antigravity and gravitational energy
18. Article 250: Planet X causing the earth to move away from the Sun
19. Article 251: Active Object in inner Solar System tracked by the Stereo A Spacecraft Hi2 detectors
20. Article 252: Planet X cosmic rays and human health
21. Article 253: Is the earth flat?

Chapter 1

Introduction

This is Part II of a series containing so far unpublished Articles I wrote after writing my first 4 books. Part I ended with Article 232 and thus Part II starts with Article 233. The Articles cover several subjects from gravity theory to the effects of the Planet X system on the Earth, the Sun, and the Solar System. They also cover simulation devices and other deception methods and much more.

Dr. Claudia Albers

Planet X physicist

July 4[th,] 2018

Chapter 2

233: The holographic Sun in the clouds

In the satellite image below there is an area of cloud that looks bright orange, and it thus seems that a light source is present inside the cloud, in that area of the atmosphere. This light source would be one of the Sun simulation models that are in operation in the earth's atmosphere (see Article 226b: Sun simulating devices: the irrefutable evidence) [1].

Figure 2.1. A bright orange region, below clouds, on the top left part of the image, indicates the presence of a Sun simulator, inside those clouds. The shadows formed by clouds indicate that the atmosphere is opaque, at the altitude, at which these clouds have formed. It is not normal for the atmosphere to be so opaque, at this altitude, and it is therefore likely that this was artificially achieved through aerosol spraying. This type of effect is usually seen in heavy fog conditions, which is only normal close to the ground (see Article 229: Chemtrail effects: Earth's atmosphere is now opaque to sunlight at cloud altitudes) [2].

One of the Sun simulating devices that has been observed in the atmosphere is the object was seen below, which has clouds in front, and behind it, and is not perfectly spherical, as we would expect the real Sun to be.

Figure 2.2. The Sun is both behind and in front of clouds, which is only possible, if the Sun is actually an artificial device, simulating the Sun, and operating at cloud altitude. The device is clearly not perfectly circular but has a jagged outline [1].

But now some video footage, taken from an airplane, actually shows what this Sun looks like, from close above the clouds, in which it is created.

Figure 2.3. Screenshots of footage, taken from an airplane, showing the Sun simulator in the clouds: This footage was shown in a Youtube video by Shahzwar Bugti [3]. The bright yellow part seems to keep appearing besides the airplane, even as it moves forward, indicating that a lens system, is in operation, in addition to, a lighting system. It seems that the cloud is being used as a three dimensional screen, for a holographic type of projection. The bright yellow part also appears to be quite small from the airplane's perspective.

In the same video showing the airplane footage, of the Sun in the clouds, the Sun simulator also seen in figure 2.2 is shown once again. This Sun has clouds in front, and behind, and is not spherical, but it seems that it is also not made out of anything solid, as a flying craft is seen flying through it.

Figure 2.4. The Sun in the clouds once again: this Sun simulator image also comes from the video showing airplane footage of the Sun in the clouds [3]. Clouds appear in front and behind the object, which is also clearly not perfectly spherical.

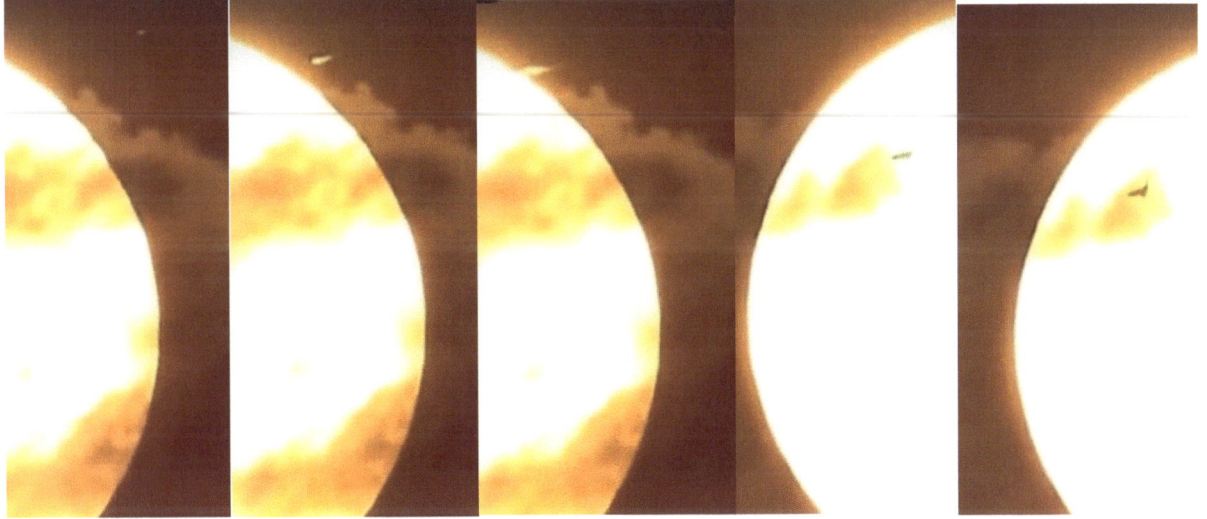

Figure 2.5. An airplane, which seems enveloped in a cloud, approaches the Sun simulator. The airplane seems to sharply increase in size, as it approaches, indicating the presence of a magnifying lens, between the Sun simulator and the camera. Once inside, the airplane is not seen, for a moment, and then begins to emerge from the simulator, at a point closer to its left side.

The fact that the airplane can fly through the Sun simulator, suggests that the simulator is not made of a solid substance, but that it is a holographic projection in the cloud. The cloud, which is most likely produced through aerosol spraying, or chemtrails, seems to act as a 3 dimensional screen. This Sun simulator seems to, therefore, be produced in the same way, as the one seen in the clouds, from the airplane, as shown in figure 2.3.

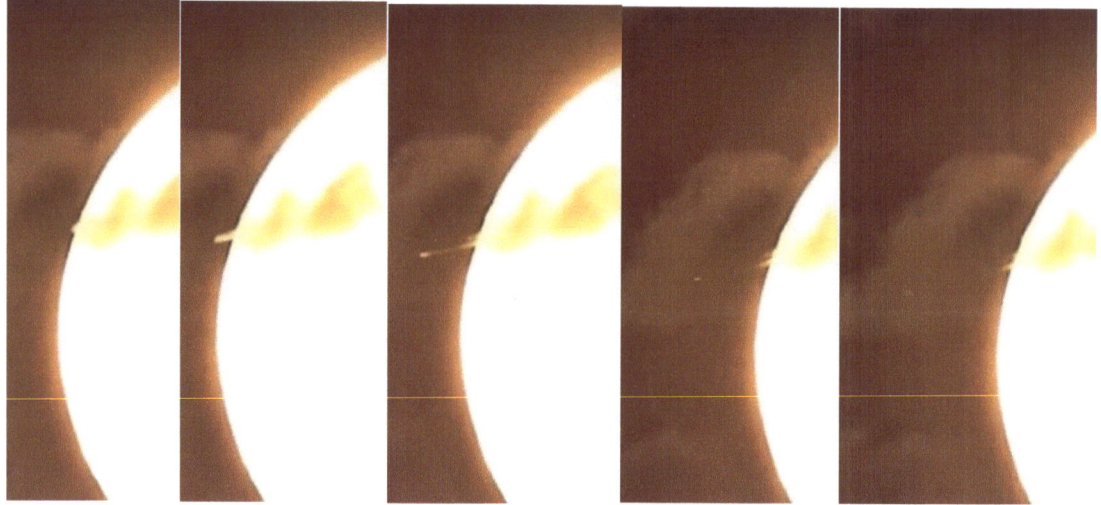

Figure 2.6. The airplane is then seen seeming to emerge from within the cloud again, showing that it is now being seen through another lens system. The airplane appears to be enveloped in the cloud at first. The cloud seems to spring back from the airplane to the simulator indicating that it has the ability to stick together. The airplane becomes much smaller as it moves away from the simulator indicating that it is no longer being magnified by the lens system.

The fact that the cloud, which is being used as a 3 dimensional screen, sticks together, is a likely indication that it contains metals, and is conductive, which allows it to be manipulated through the application of electromagnetic fields. The fact that holographic projection technology is being used to simulate the Sun in the sky agrees with the evidence that technology appears to be a type of projector has been photographed in the sky (see Article 231: Advanced technology in the sky) [4].

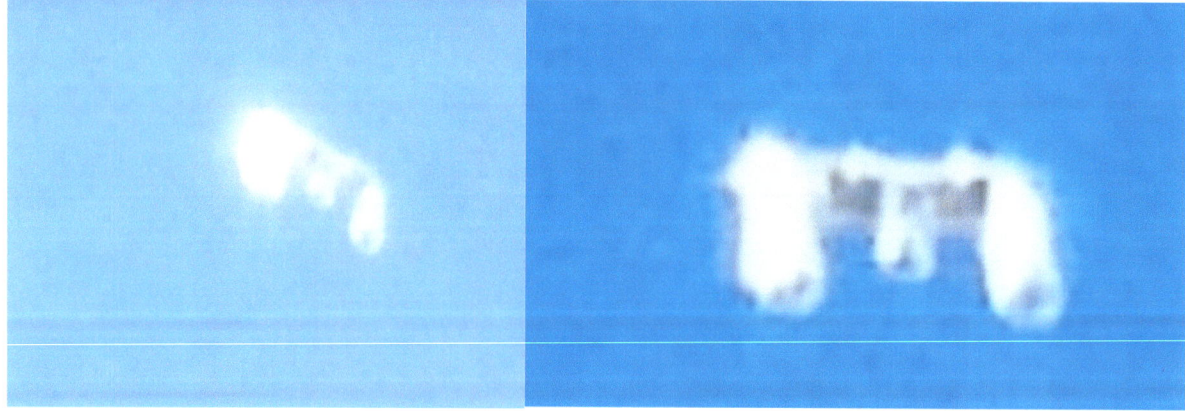

Figure 2.7. Artificial device with what seems like projectors and screens on it, suggestive of what a holographic projecting device may look like [4].

In conclusion, the Sun simulator, shown in figure 2.2, does not seem to be made of any solid substance, but seems to be a hologram, projected onto cloud, which appears to be conductive, and thus artificially created, through chemtrails, and which can thus be manipulated, through the application of electromagnetic fields. The cloud acts as a 3 dimensional screen, onto which the hologram is projected.

References:

[1] Albers, C. (2018). Article 226b: Sun simulating devices: the irrefutable evidence.
[2] Albers, C. (2018). Article 229: Chemtrail effects: Earth's atmosphere is now opaque to sunlight at cloud altitudes.
[3] https://www.youtube.com/watch?v=eN8yhUzqLCA&list=LLURC-iCzVpz1Jl010tIaZ8Q&index=2&t=0s
[4] Albers, C. (2018). Article 231: Advanced technology in the sky.

Chapter 3

234: Stellar Core in LASCO C3 image has a comet like characteristics

Figure 3.1 shows a faint object in a LASCO C3 image, from May 20th, 2018. The object's angular width is about 2.5 times that of the Sun. The Sun's size is indicated by the white circle on the coronagraph occulter. The object seems to have developed comet characteristics, namely a coma and two comet-like tails, although these are much fainter than what is usually seen in comets.

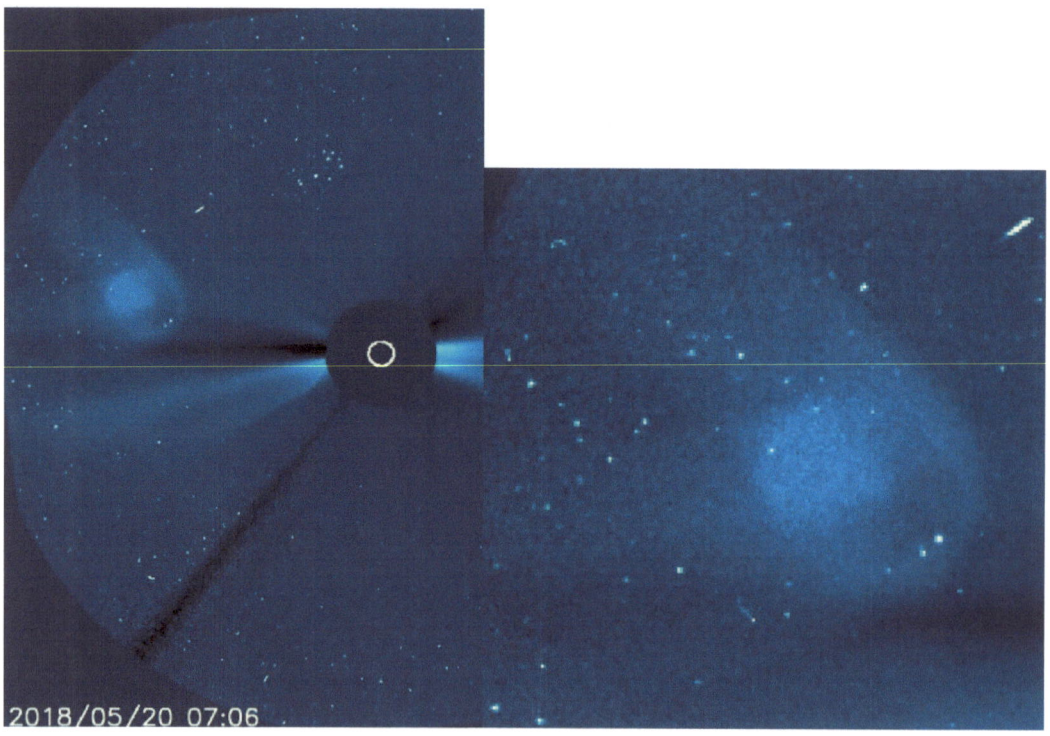

Figure 3.1. LASCO C3 image in which a large white spherical cloud appears followed by what appears to be trails. The trails have to tail since the object could not have travelled along 2 different directions. Tails are due to material moving toward the object, from behind it, and meeting electrons coming from the Sun (see Article 170: Comets, planets and crustal displacements) [1]. The object has a smaller dark circle in it which forms its own dark trail, within the white diffuse trails. The dark trail indicates that less light is emitted from this area. The dark circle seems to, therefore, be the nucleus of this comet-like object and therefore indicate its original size.

The faint trails, behind the object, seem to be comet-like tails; these will be due to the positively charged material, drawn from one of the Sun's nebular clouds. The material is drawn in toward the object, whilst a current of electrons is drawn from the Sun's corona. The two currents meet at the object's surface and give rise to a cloud of material, which emits light around the object, and which is called a coma. The coma may be many times larger than the actual comet nucleus. In this case, it appears to be only 2.5 times larger. The electrons will also combine with the positive ions in the tail. Whenever a positive ion captures electrons, photons are emitted. It is these photons that cause both the coma and the tail to give off light (see Article 170: Comets, planets and crustal displacements) [1]. The dark circle indicates the size of the original object and the fact that it is absorbing electrons, coming from the Sun, leaving fewer electrons to combine with the positive ions coming from the Sun's nebular clouds and thus producing both a surface and a region, within the tails, from which photons are not emitted.

Figure 3.2. A bright comet with two bright tails streaks across the sky.

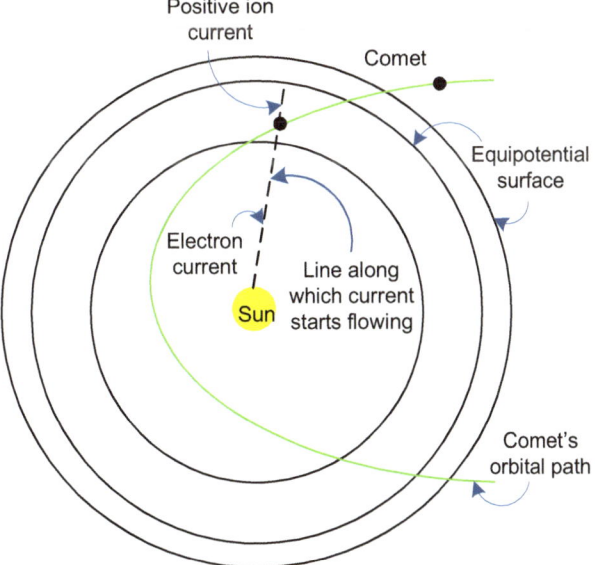

Figure 3.3. Comets, like all other objects in the universe, have a negative outer layer, which draws a current in the solar capacitor, ions from a nebular cloud and electrons from the sun's corona. The more elliptical the orbit, the more current, it will draw, and the brighter it will be.

The object, in the LASCO C3 image above, is not as bright as normal comets suggesting that it does not draw as strong a current as a comet would. This indicates that it does not have a normal outer negative layer of electrons, which suggests that it is a Stellar Core, since these objects are known to be depleted in electrons, and act as super ions, whereas normal celestial objects, such as the Sun and its planets, act as super atoms (see Article 193: Stellar Cores in the Sun's corona: why do they not collide with the Sun?) [2]

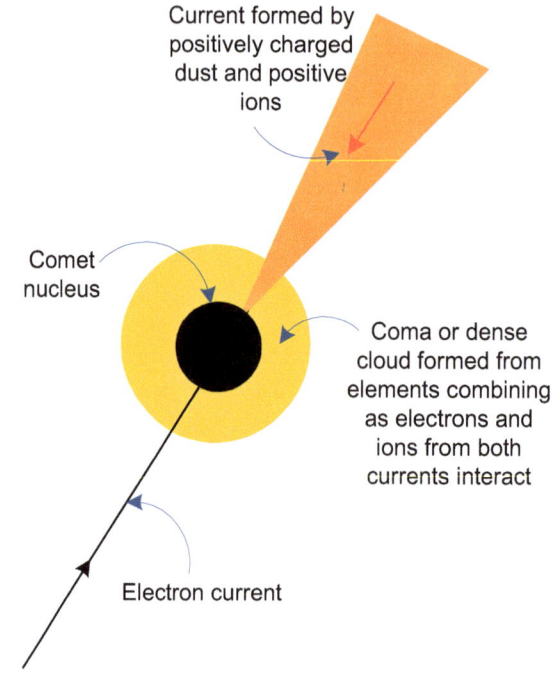

Figure 3.4. An electron current flows from the Sun, and a current of positive ions and positively charged dust flows from one of the nebular ion clouds, inwards, towards the Sun, along with a line passing through the comet. A coma or dense cloud forms around the comet, which is made of elements formed as a result of the interactions between the neutral atoms formed. Hydrogen and water will combine into water (see Article 169: Planetary formation: comets to planets) [3].

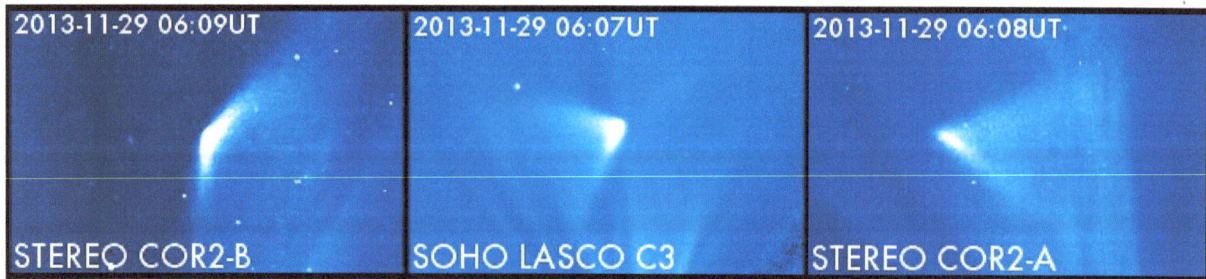

Figure 3.5. Comet ISON in Stereo COR2 B and LASCO C3 images produces tails, behind it, similar to what can be seen behind the object in figure 1. No spherically shaped coma is apparent, however, as comets

are not usually spherical. But the object in figure 1 produces a perfectly spherical coma, suggesting that this particular object is perfectly spherical and very large.

This is not the first time that this type of object has been seen in the LASCO C3 images, very similar objects have been observed in 2009 and 2016 as shown below.

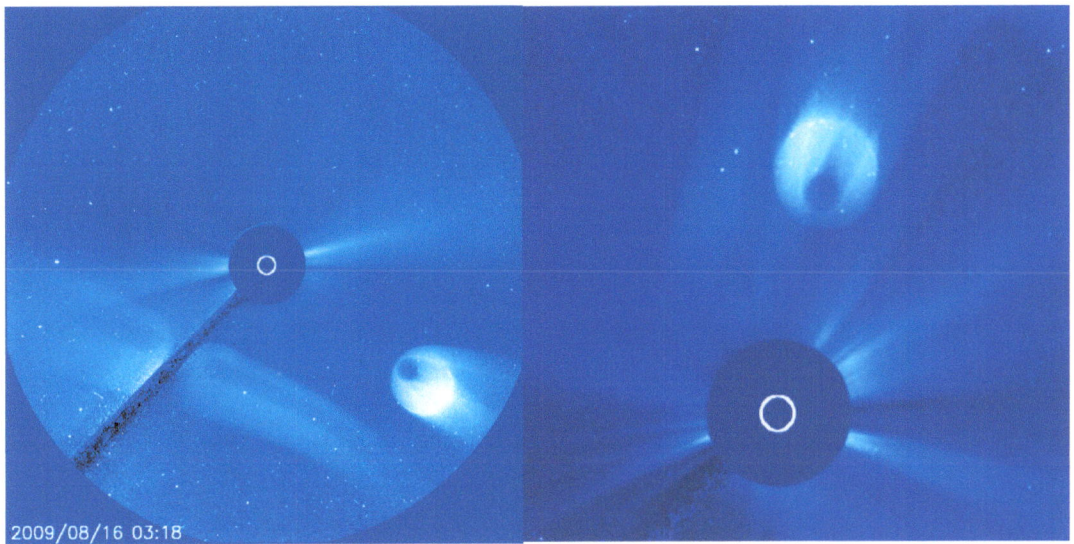

Figure 3.6. Objects followed by white tails, with a dark trail, inside the wider white one. The dark region within the tail is about the same width as the object's diameter, and thus indicates a region, where fewer electrons are available, to combine with ions, and therefore a region where fewer photons are given off because the object absorbs electrons. The image on the right is from February 14^{th,} 2016. One of the objects seems brighter than the object seen in the 2018 image, suggesting that it draws a stronger ion current and is thus not as depleted in electrons as the 2018 Stellar Core in figure 1. The two fainter tracks in the 2009 image do not align with the Sun which suggests that the two objects are drawing electrons from another object located between the two.

An object, which absorbs electrons, will be depleted in electrons, which the Stellar Cores that have invaded the Solar System are known to be. The object in figure 1 and the objects in the above images, therefore, appear to be more objects belonging to this system of invading dead stars, which come to the Sun and absorb energy from it.

Figure 3.7: SDO image in the 171 angstrom wavelength from October 13th, 2017 showing a dark Stellar Core, which appears to be about half of the size of Jupiter, drawing matter from the Sun.

Now, comet tails always point away from the Sun, and according to James McCanney are very narrow when the comet is far from the Sun, and become very wide when the comet approaches perihelion position [4]. From the comet tail orientation, we can determine on which side of the Sun, the large object in figure 1, is.

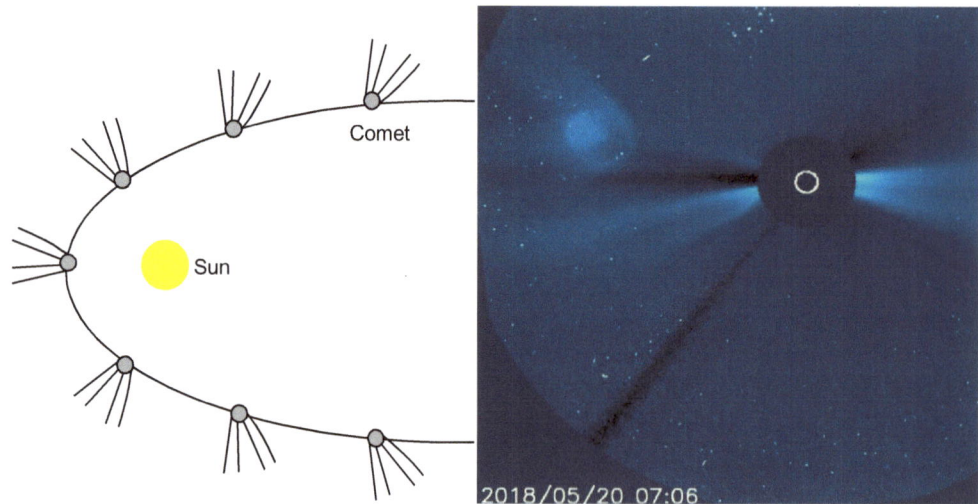

Figure 3.8. The object of interest has tails which point in a direction (toward the left and slightly upward), which suggests that it is very close to the Sun and perhaps just behind the Sun, in the LASCO C3 image.

The object of interest, therefore, appears to be very close to the Sun, and therefore its size may be estimated, from a direct size comparison with the Sun. The Sun's size is indicated by the white circle, on the occulter. The coma or white spherical part is about 2.5 times the size of the Sun, but the dark circle or original Stellar Core inside the coma is about the same size as the Sun.

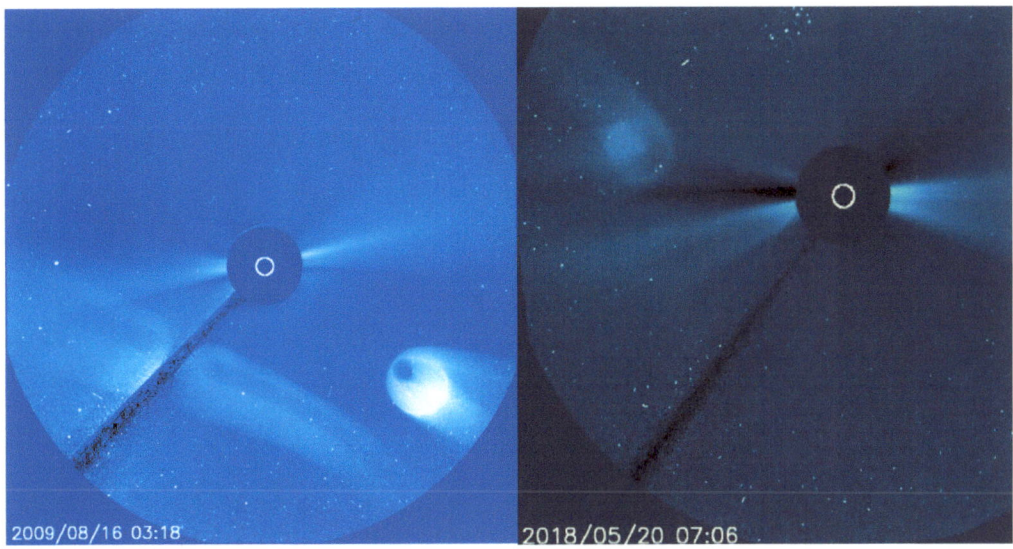

Figure 3.9. Stellar Cores in LASCO C3 images from 2009 and 2018. The fainter tracks indicate a more depleted state of electrons and thus a Stellar Core which is less able to draw a current in the solar capacitor.

One of the objects seen in 2009 and 2016 is brighter than the one in 2018 but the others are fainter, suggesting that the Stellar Cores associated with these fainter tracks were more depleted in electrons than the Stellar Core producing the brighter track. Stellar Cores as explained in Article 184: Stellar Core evolution [5], arrive in the Solar System so depleted in electrons that they have a neutral rather than a negative outer layer.

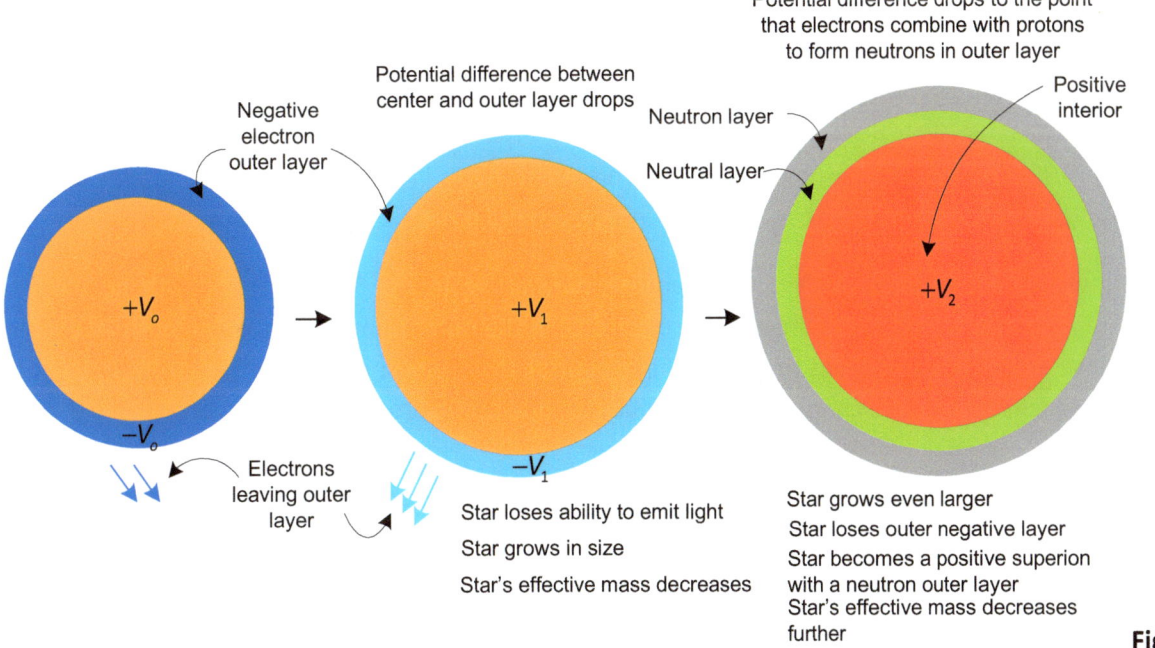

Figure 3.10. Illustration of how a star's ageing process, which causes a decrease in gravitational energy, is likely to turn it into a Stellar Core. At the Stellar Core stage (neutron outer layer) any electron captured

by the star would, most likely, increase the size of the neutral layer, rather than form a negative outer layer, so the star would remain deficient in electrons until it enters the Sun's corona.

It is, therefore, the positively charged inner core that causes them to be attracted to the Sun's outer layer and it is thus the electrostatic force that brings them in.

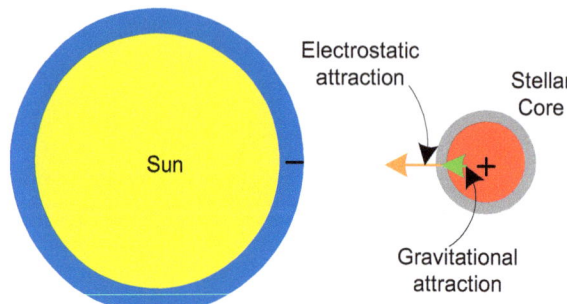

Figure 3.11. A Stellar Core is a dead star that has lost the ability to generate gravitational energy in the core and has become severely depleted in electrons. The object is electrostatically attracted to the Sun. The gravitational attraction between the two is very weak

In their initial state of electron depletion, Stellar Cores will not draw a current from the solar capacitor. But once they make contact with the Sun's corona, they will draw electrons, from the Sun, and start replenishing their supply of electrons, whilst cause the Sun to become depleted. Initially, the electrons they will draw, will move below the surface, toward the inner core, but eventually as the object gains gravitational energy, by drawing matter from the Sun containing protons, it will be able to start forming a negative outer layer of electrons. Once this negative outer layer is in place, the Stellar Cores will be able to draw current, from the solar capacitor, and will thus produce comas, and comet type tails, as seen in the image, in figure 1. The fact that it produces a dark trail, indicates that it is still absorbing electrons and that the process of absorbing enough electrons, may take a very long time. In addition, since they will most likely draw electrons, from each other, as well as the Sun, it is possible this process will never be complete, as new objects seem to continue arriving in the Solar System, and they will thus continue to draw electrons from the Sun, indefinitely, or until it has no more to give.

Comets would not produce a dark trail because they arrive in the Solar System, with no electron depletion, in other words, there are enough electrons, in the outer layer, to cancel the positive charges on the inside, which is analogous to them being a neutral atom, which has enough electrons in its outer shells, to cancel the positive charge, inside the nucleus, due to the protons present there.

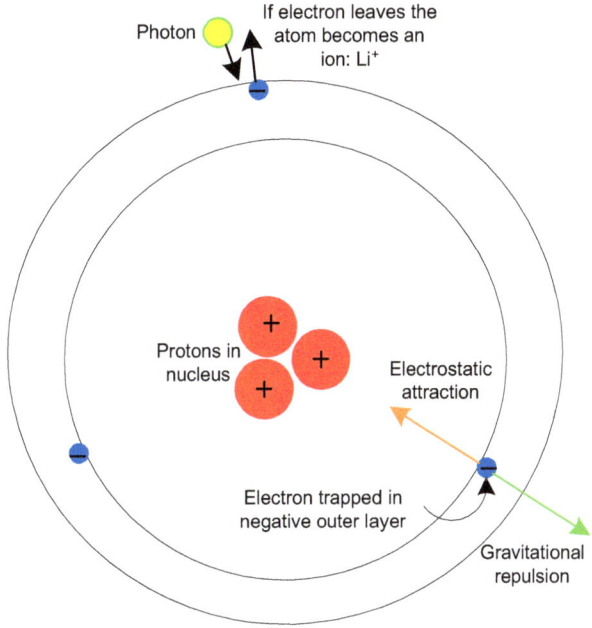

Figure 3.12. Lithium atom: the atom is neutral, when there are as many electrons, in the outer orbits, as protons, in the nucleus. The atom becomes a positive ion, once it loses one, or more, electrons. Normal stars, planets, and asteroids, or comets, act like super neutral atoms, but Stellar Cores act like super ions.

In conclusion, the object seen in a LASCO C3 image appears to be a Stellar Core, which although still depleted in electrons, has been in the Solar System long enough to have developed an outer layer of electrons, so that it is now able to draw a current, within the solar capacitor, and thus produce comet characteristics, namely a coma and tails. The dark track indicates that the Stellar Core is absorbing electrons coming from the Sun's corona, and thus still in a depleted state. In addition, the object must be following an elliptical orbit, as objects in circular orbits will not draw current from the solar capacitor.

References:

[1]   Albers, C. (2018). Article 170: Comets, planets and crustal displacements.
[2]   Albers, C. (2018). Article 193: Stellar Cores in the Sun's corona: why do they not collide with the Sun?
[3]   Albers, C. (2018). Article 169: Planetary formation: comets to planets.
[4]   McCanney, J. (2002). Planet X Comets and Earth Changes. Jmccanneyscience.com press Minneapolis.
[5]   Albers, C. (2018). Article 184: Stellar Core evolution.

Chapter 4

235: Is there a Planet 9 in the outer reaches of our Solar System?

It has recently been widely publicized that an extreme TNO (Trans Neptunian Object) with a strange orbit, suggests that the Planet 9 hypothesis may provide an explanation for its extreme orbit. The object is called 2015 BP519 and was discovered 3 years ago. The object appears to be in highly inclined, and extremely eccentric, orbit. In fact, the eccentricity of the orbit is 0.92, which means that it is getting close to being open, or a hyperbolic orbit. In addition, the object's orbital inclination is 54° [1].

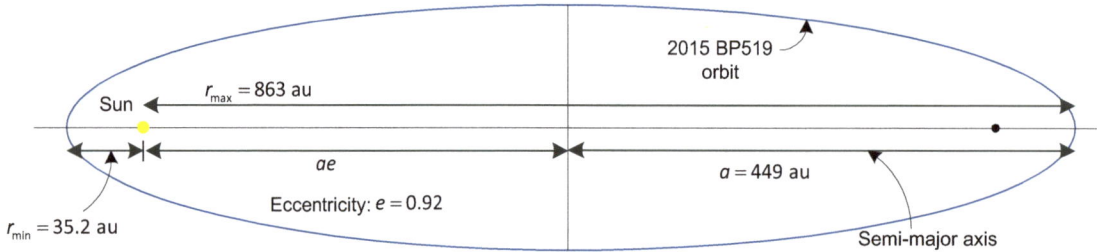

Figure 4.1. The asteroid 2015 BP519 has a highly eccentric orbit as its orbital eccentricity is 0.92. An orbital eccentricity above 1, corresponds to an open, or a hyperbolic orbit. The object never comes any closer than Neptune's orbit (30 au) as its perihelion position is 35.2 au. Its average distance, from the Sun, or semi-major axis, is 449 au, where au stands for astronomical units and 1 au is the distance between the Sun and the Earth. The Solar System is believed to have a radius of about 100 au.

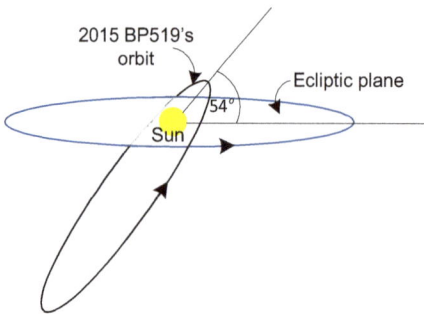

Figure 4.2. Illustration of 2015 BP519's orbital inclination, which is a high 54°. The average inclination of TNOs (Trans Neptunian Object) or objects whose average distance from the Sun is greater than 30 au, is 17.3°, so this object's inclination is high.

Because this object's average distance from the Sun is much greater than Neptune's orbit, it is called an extreme TNO (Trans Neptunian Object). Even though this object's orbital inclination is way above the average, it is not as high as another TNO called Niku. Niku, classified as a minor planet, has a near polar orbit, and in addition, it has a retrograde orbit, which suggests that it is a new acquisition to the Solar

System (see Article 226: Niku recently discovered newcomer in the Solar System) [2]. Objects that have been a part of the Solar System, for billions of years, are likely to orbit in the same direction as all the planets, in the solar system, and also to have a nearly circular orbit, as an elliptical orbit causes the object to draw current, in the solar capacitor. This has a drag effect, on the object, which thus circularizes its orbit (see Article 169: Planetary formation: comets to planets) [3].

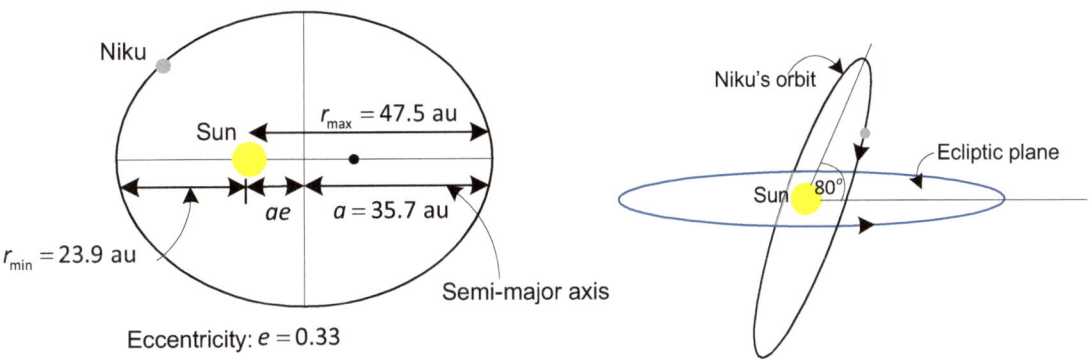

Figure 4.3. Niku is a minor planet and TNO with a near polar retrograde orbit. The eccentricity of its orbit is 0.33. The fact that the orbit is so extremely inclined and retrograde suggests that this is a new addition to the Solar System and the Solar System is a dynamic system, where new objects are constantly added [2].

According to Becker, et al. [1] who analyzed 2015 BP519's orbit, it is not possible to explain the extreme tilt of the orbit, unless another planet exists beyond the orbit of Neptune. However, if this object, like Niku, is a new acquisition of the Solar System, there is no need to explain the evolution of its orbit, in terms of known solar system dynamics, simply because there has not been enough time, for such an evolution.

Astronomers have assumed that the Solar System is a static system that never changes, but the fact is that it is a dynamic system that is constantly changing, with new objects coming into it, all the time. The fact that a System of dead stars or Stellar Cores have been invading the Solar System, for at least 150 years, is evidence of that (see Article 146: Planet X System: time of arrival) [4]. Some of the evidence showing that these objects have been observed in the Sun's corona is shown below.

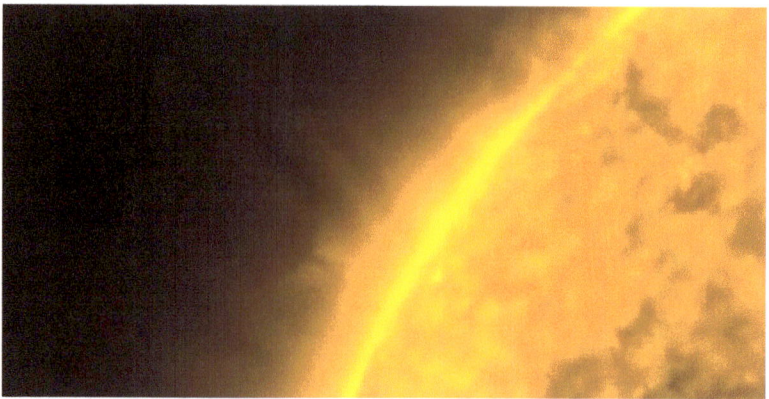

Figure 4.4. SDO image in the 171 angstrom wavelength from October 13th, 2017 showing a dark Stellar Core, which appears to be about half of the size of Jupiter, making a vortex connection with the Sun.

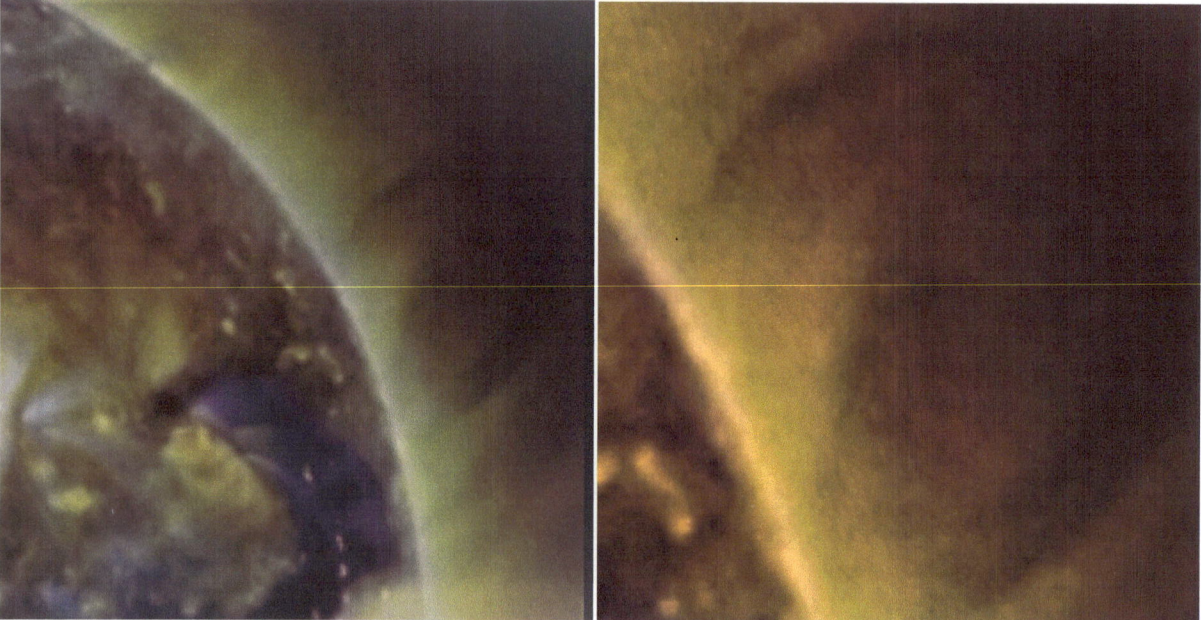

Figure 4.5. Planet X Object, or Stellar Core, appears in composite SDO images, from December 25th, 2017. This object is about one fifth of the size of the Sun, or twice as large as Jupiter. Surface features looking like craters and layered material, with well-defined outlines, are clearly visible showing that the object is emitting ultraviolet light.

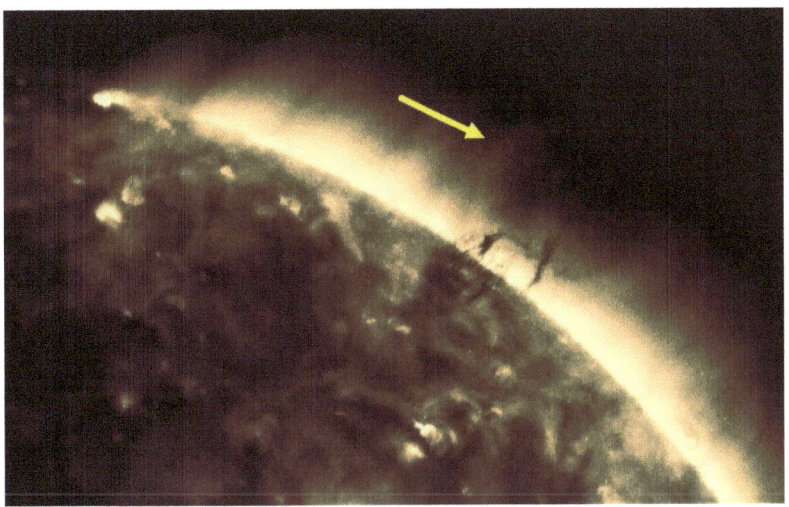

Figure 4.6. SDO image in the 193 angstroms wavelength: a Stellar Core, in the Sun's corona, making a matter connection with the Sun.

These objects have been draining the Sun, for many years now, and the Sun is weakening fast, as a result. For more details on this see Article 195: Stellar Cores and the dying Sun [5] and Article 225: Weakening Sun: SORCE radiation measurements are not all solar radiation [6]. The end result is that the Sun will go dark just as its twin, Nemesis, may already have done, as a result of being invaded by members of the same system of Stellar Cores (see Article 208: Incoming Dark Star) [7].

It is also widely believed that comets, which have very eccentric orbits, come from the Oort Cloud, or a shell shaped cloud of small objects, far beyond the orbit of Pluto. However, there is actually no evidence that such a cloud exists, and it is more likely that comets just come into the Solar System from outside it. In other words, it is likely that comets are interstellar travellers and that therefore all objects, with extremely eccentric, or inclined, orbits, such as Niku and 2015 BP519, are new acquisitions, which recently entered the Solar System, and became newly captured objects.

The evidence that the System of dead stars, which have invaded the Solar System, are orbiting and affecting the earth, lies in the unprecedented tidal events, which have occurred along many coastlines around the world. Only an object exerting a small gravitational attraction, coming very close to the earth, would be able to exert a tidal force, leading to such an effect, and since the Stellar Core's gravitational attraction is weak, this is the exact effect we would expect (see Article 227: Stellar Cores affecting earth and possible connection to Volcanic Eruptions) [8].

Figure 4.7. Ocean recedes leaving boats sitting on mud, in the harbor in Punta del Este, Uruguay, on August 11th, 2017. The ocean came back but this extreme low tide had never happened before.

In conclusion, extreme TNO 2015 BP519 is much more likely to be a recent addition to the Solar System, and thus further evidence that the Solar System is a dynamic system, with new objects being constantly added, rather than being evidence that another planet, beyond the orbit of Neptune, is influencing its orbit. We have evidence that extremely large objects are in the inner solar system and are affecting the Sun and the Earth. These objects are dead stars or Stellar Cores, and they have the potential of deeply affecting life on our planet. These objects should, therefore, be the real subject of interest, not a planet in the outer reaches of the Solar System that cannot, in any way, affect our planet. However, it seems that the planet 9 hypothesis is actually used as a diversion, in order to focus people's attention away from the real truth.

References:

[1] Becker, J. et al. (2018). Discovery and Dynamical Analysis of an extreme Trans-Neptunian Object with a High Orbital Inclination. https://arxiv.org/pdf/1805.05355.pdf
[2] Albers, C. (2018). Article 226: Niku recently discovered a newcomer in the Solar System.
[3] Albers, C. (2018). Article 169: Planetary formation: comets to planets.
[4] Albers, C. (2018). Article 146: Planet X System: time of arrival.
[5] Albers, C. (2018). Article 195: Stellar Cores and the dying Sun.
[6] Albers, C. (2018). Article 225: Weakening Sun: SORCE radiation measurements are not all solar radiation.
[7] Albers, C. (2018). Article 208: Incoming dark Star.
[8] Albers, C. (2018). Article 227: Stellar Cores affecting the earth and possible connection to Volcanic Eruptions.

Chapter 5

236: What are Stellar Cores?

Stellar Cores are celestial objects belonging to a large system of objects that have invaded our Solar System. These objects have been extensively observed in the Sun's corona, and often in the process of making a matter connection, with the Sun. This connection is in the shape of a vortex. Their gravitational interaction, with the Sun, and other celestial objects, is weak, so that they may be as large as the Sun, but interact as if they have the mass of a very small moon. The interaction through which they draw matter from the Sun is tidal in nature. They are however powerful objects with a very powerful ability to draw energy from the Sun. They do not emit visible light when they first arrive in the Sun's corona; they may, however, emit infrared radiation. But after drawing energy from the Sun, for some time, they are able to emit radiation in several wavelengths, namely x-ray, ultraviolet and visible light. They are, however, never brighter than the Sun as the energy they gain, which then allows them to emit light, comes from the Sun. The fact that they are x-rayed sources [1] is one of the ways that it can be deduced that they are stars. They are stars that once emitted light, like our Sun, but have run out of energy, and died. They are dead stars; they are dead because they can no longer generate energy in their cores. They can regenerate up to a point, by drawing energy that the Sun generates, from the Sun, or other planets, in the Solar System, but they cannot regain the ability to generate their own energy. This means that they remain dependent, on their host body, and are thus never likely to withdraw from it.

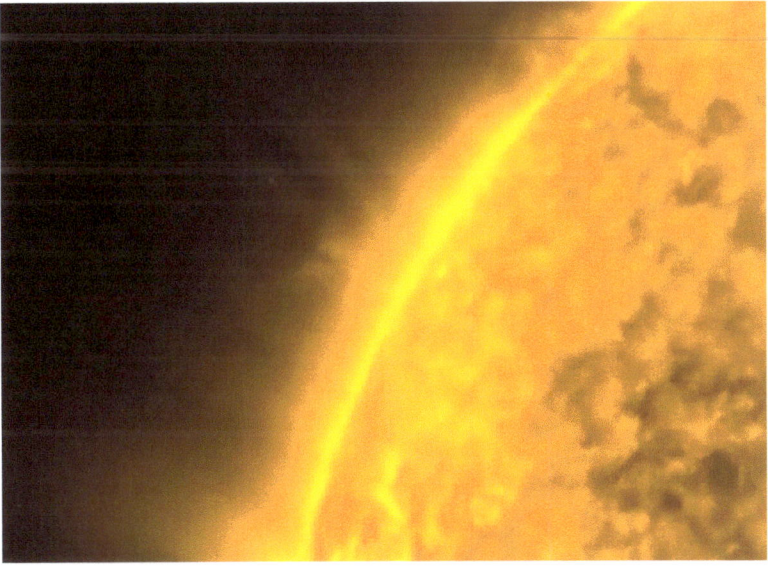

Figure 5.1. SDO image in the 171 angstrom wavelength from October 13th, 2017 showing a dark Stellar Core, which appears to be about half of the size of Jupiter, making a vortex connection with the Sun. This object is not yet able to emit ultraviolet light, in this wavelength, and thus looks black in this image.

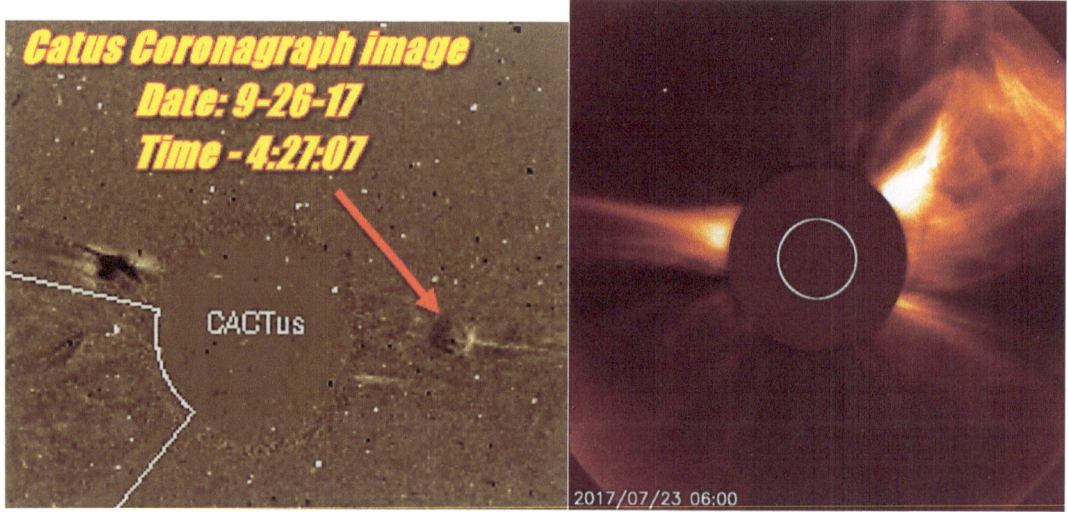

Figure 5.2. Coronagraph images showing Stellar Cores, in the Sun's corona, during CMEs. The objects seem to provoke the Sun into having CMEs and solar flares. The Stellar Core in the right image has to be very close to the Sun, in order to be enveloped in the CME material, right after it was ejected by the Sun, and therefore its size may be estimated by comparing it to the Sun. The object is about the same size as the Sun. The white specks are part of the debris around the Sun due to the material the Stellar Cores shed.

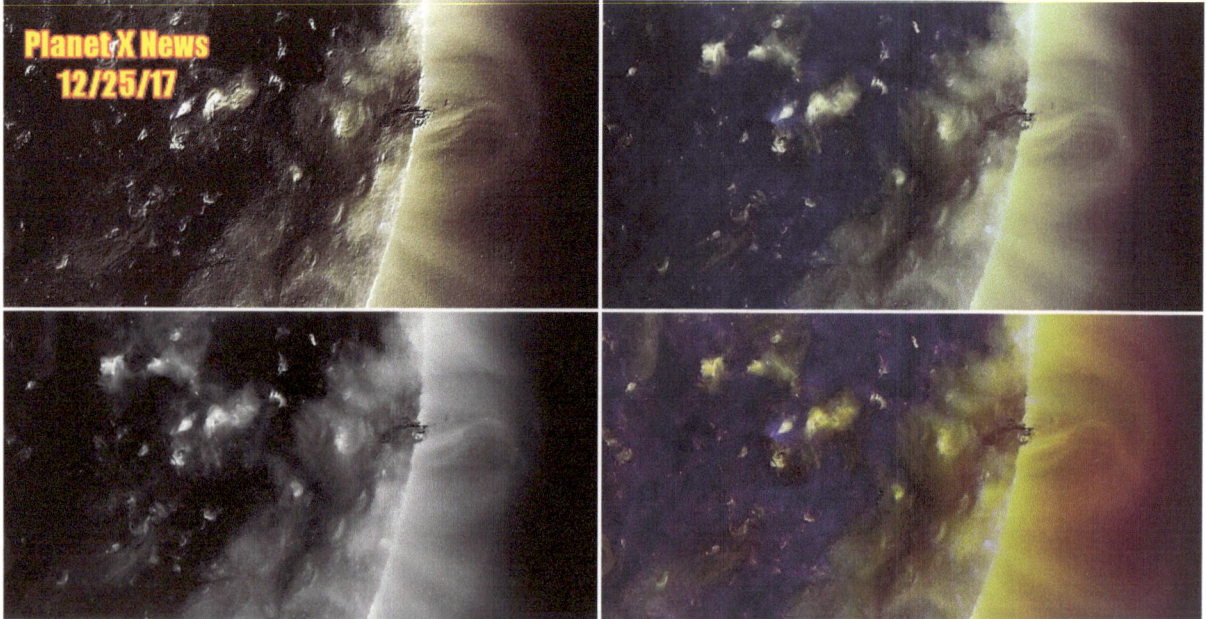

Figure 5.3. SDO composite image showing a Stellar Core, in the Sun's corona, in various different shades: It is obviously emitting ultraviolet light or it would not be seen in the image's computer generated color. The object is striped and a size comparison with the Sun reveals that it is about 4 times larger than the earth. The trail behind it indicates that it is drawing material from the Sun's corona.

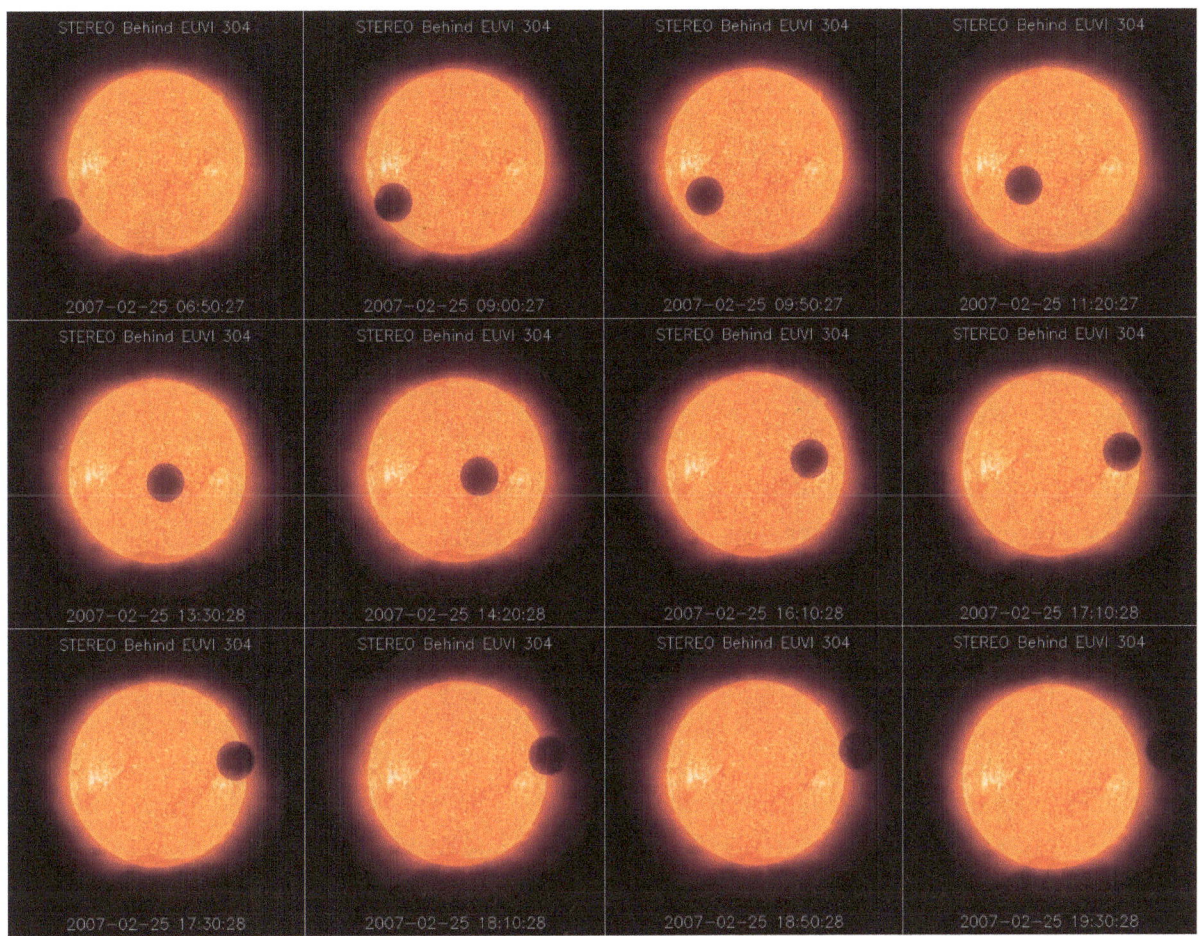

Figure 5.4. Stereo B EUVI 304 angstrom wavelength image from 2007: A large object traverses the Sun. The object is not black against the background of the solar surface, indicating that it is in the Sun's corona. The object is 2.2 times the size of Jupiter, takes 10 hours to traverse the Sun and is travelling at 39 km/s or 24 mi/s, or at a much lower speed than the Sun's escape velocity of 616 km/s. This indicates that it has a very weakened capacity to gravitationally interact. (For more detail see Article 153: Planet X: Escape velocity and Gravity [2] and Article 210: Stellar Core gravity: tidal and G is not constant [3].

The name Stellar Core comes from the observation that these dead stars have solid surfaces and that even the material clinging to the surface, which gives rise to their often striped appearance, appears to be solid, but clumpy. These observed surfaces seem to be what is left of a star, and it looks like a dense and solid core. The Stellar Cores shed the material, making up their outer layers, when they arrive in the Sun's corona, and appear to then gain a new outer layer and atmosphere, from the Sun's corona, or from whatever another object that becomes their host. There is evidence that the earth is also hosting some of these objects (see Article 227: Stellar Cores affecting the earth and possible connection to volcanic eruptions) [4].

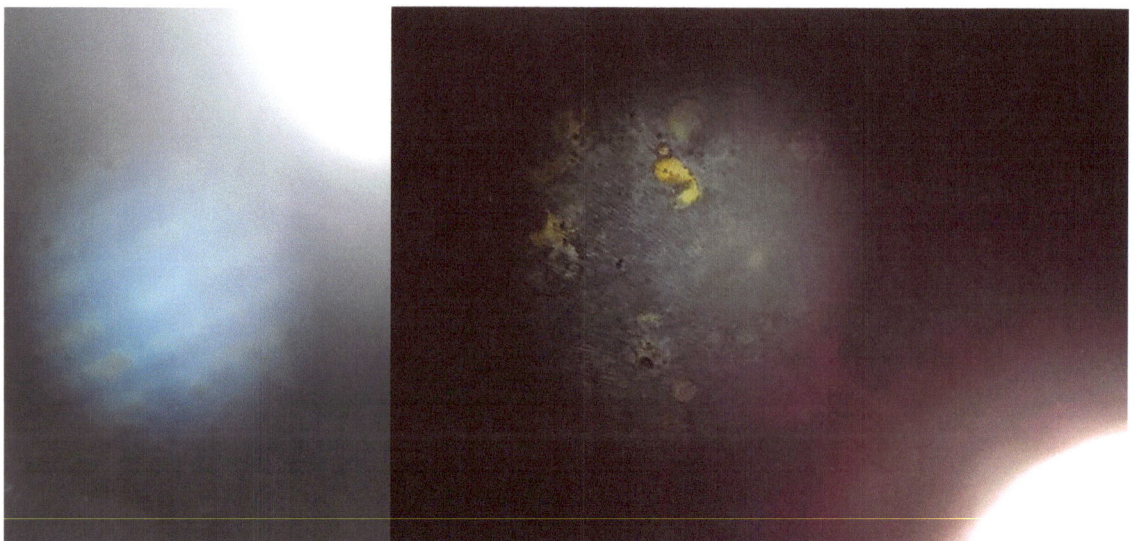

Figure 5.5. The Blue Stellar Core photographed, in the Sun's corona, through a telescope, in May 2017 (left) and in July of 2017 (right). The object shed much of the material that made up its stripes, seen in the earlier image, in the ensuing time. The surface is blue because it emits blue light. The surface has grooves in it, and is therefore solid, the yellowish material, clinging to it, is also in the solid state. The object seems to be building up a gaseous atmosphere, which emits magenta (pink) light. This gaseous material must be ionized, or in the plasma state, in order to give off this color of light. The surface gets progressively lighter in color from one circular spot.

The debris, formed from the Stellar Cores shedding their old outer layers, is seen floating around them, and the Sun. This material is not attracted toward the Sun, or the Stellar Cores, indicating that the strength of the gravitational attraction between it and other objects is zero or very close to zero. This is further evidence that the gravitational interaction is variable in nature, and connected to the energy an object has or is able to generate. The Stellar Cores in time gain energy which strengthens their ability to interact gravitationally but this energy comes from their host body.

The Sun has been severely weakened as a result of the Stellar Cores interaction with it. The Sun not only goes completely dark periodically, it is also getting progressively weaker, and less able to emit light, as well as, produce a solar wind. This can be seen in the fact that the Sun's corona is ever growing smaller, and darker, and has been doing so for many years now, and also in the fact that the Sun's magnetic field, associated to sunspots, has been progressively decreasing independent of the Solar cycle. For more details see Article 195: Stellar Cores and the dying Sun [5].

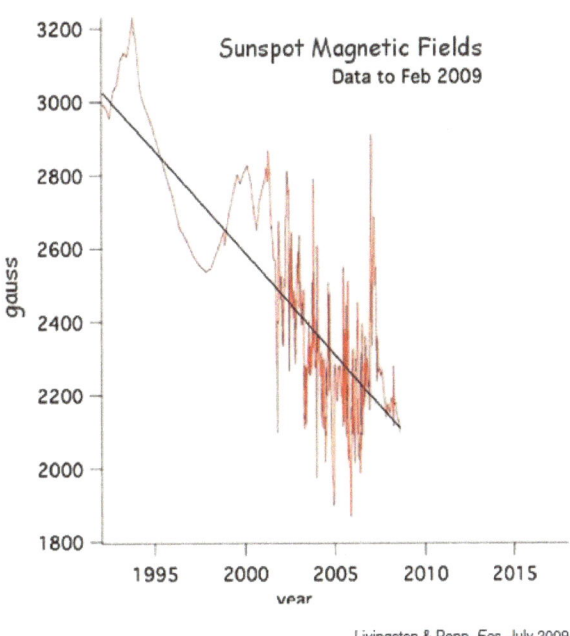

Figure 5.6. The Sun's magnetic field strength, associated with sunspots, dropped independently of the solar cycle during cycle 23. Livingston and Penn found that it consistently dropped by 50 gausses per year between 1996 and 2009 [6].

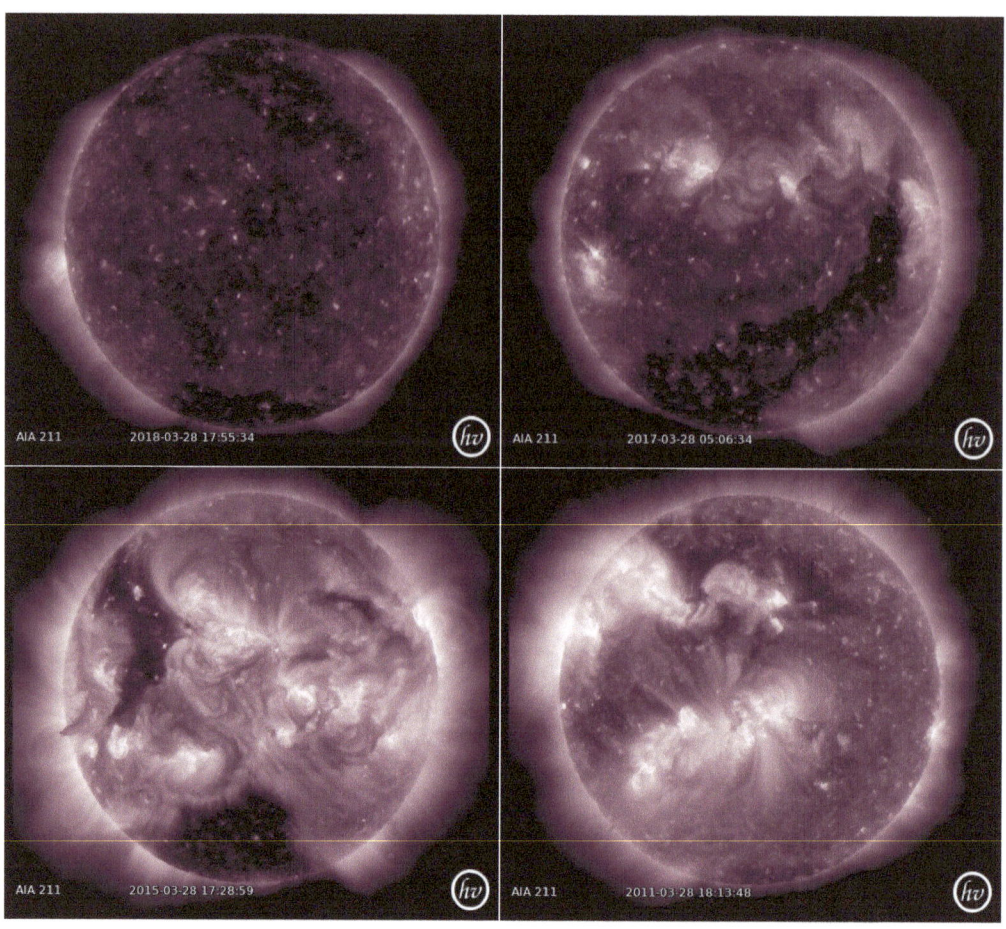

Figure 5.7. The Sun's corona in 2018 is much darker and smaller than all other years shown, including in 2011, the year closest to last solar minimum. It looks like the whole solar surface is now one coronal hole, indicating that the particle density, in the sun's atmosphere, is now extremely low.

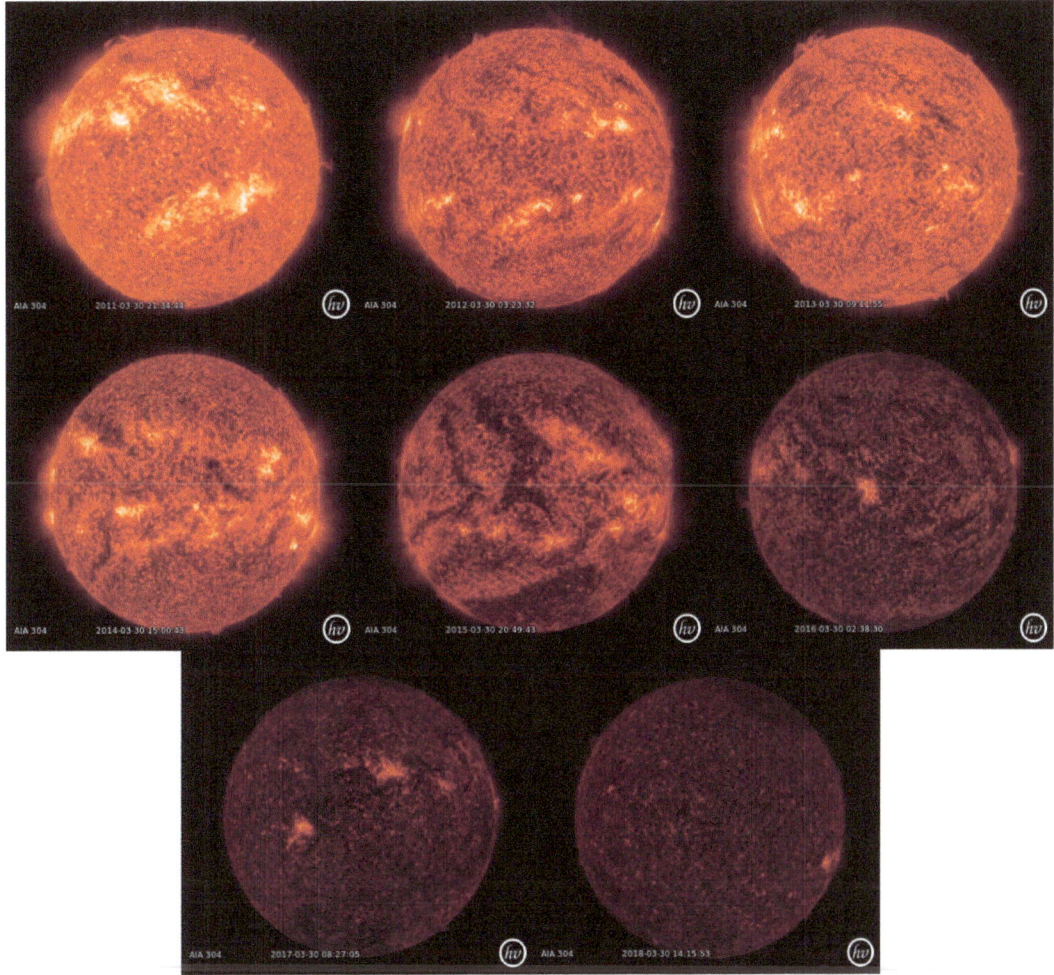

Figure 58. SDO images of the Sun in the 304 angstrom wavelength, from 2011 to 2018. The Sun's has darkened dramatically, showing that it has been drastically weakened.

Observation of these objects has led to an understanding of how gravity works (see Book: Planet X revealed Gravity and Light or articles 181 and 182) [7, 8, 9], but also that the main difference between stars and planets is size, both stars and planets have solid cores, and both generate energy, in their cores, from the decay of unstable nuclei, or fission. This energy is used to generate an electromagnetic field, and larger objects can generate a larger field, which gives rise to a greater degree of ionization and thus to more emission of radiation, than smaller ones. Because this energy is directly related to the gravitational field generated by the objects, I have called it gravitational energy. It is the gravitational interaction resulting from the object's energy status that gives rise to charge separation so that objects have positively charged cores, and negative outer layers, which then results in the generation of the electromagnetic field, which in turn, allows light emission.

In conclusion, Stellar Cores are dead stars, or stars that have lost the ability to generate energy, in their cores, and that have revealed the true nature of gravity and its relationship to the electromagnetic interaction, as well as the fact that stars and planets, are formed, and operate through exactly the same mechanism, and that both have solid cores and atmospheres, in which light emission processes occur.

These objects are destroying the Sun's ability to emit light and are also affecting the Earth (see Article 201: Africa breaking up: a preview of what is to come) [10].

References:

[1] Albers, C. (2018). Article 225: Weakening Sun: SORCE radiation measurements are not all solar radiation.
[2] Albers, C. (2018). Article 153: Planet X: Escape velocity and Gravity.
[3] Albers, C. (2018). Article 210: Stellar Core gravity: tidal and G is not constant.
[4] Albers, C. (2018). Article 227: Stellar Cores affecting the earth and possible connection to volcanic eruptions.
[5] Albers, C. (2018). Article 195: Stellar Cores and the dying Sun.
[6] Matthew J. Penn and William Livingston, "Long-term Evolution of Sunspot Magnetic Fields." http://www.probeinternational.org/Livingston-penn-2010.pdf
[7] Albers, C. (2018). Book: Planet X Revealed Gravity and Light.
[8] Albers, C. (2018). Article 181: Stellar Cores and deciphering gravity
[9] Albers, C. (2018). Article 182: Einstein's dream realized: a unified field theory of electrogravitation.
[10] Albers, C. (2018). Article 201: Africa breaking up: a preview of what is to come.

Chapter 6

237: Hawaiian Island and Megatsunamis

In my previous article, Article 236: What are Stellar Cores? [1] I wrote about the System of dead stars which have invaded our Solar System, that are often observed in the Sun's corona and that is draining energy from the Sun. I have also explained how their interaction is tidal in nature, how both planets and stars generate energy in their cores from fission, and how dead stars, or these Stellar Cores, are no longer able to generate energy in their core. I have also mentioned that the earth is also host to some of these objects. Since the earth generates a lot of energy, in its core, and these objects are attracted to this energy and absorb it, it is not difficult to see that the earth would become host to Stellar Cores, just like the Sun has.

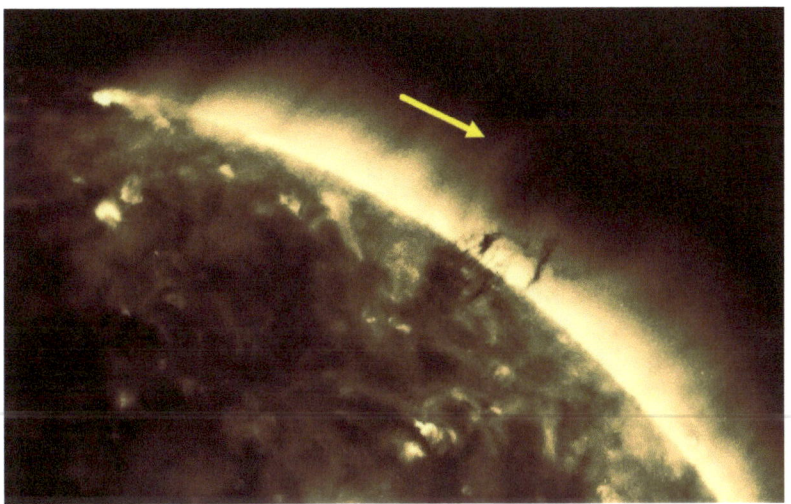

Figure 6.1. SDO image in the 193 angstroms wavelength: a Stellar Core, in the Sun's corona, making a matter connection with the Sun.

The fact that they are affecting the earth is evidenced by the unprecedented tidal events that have been witnessed, along with many coastlines worldwide. Just like the moon's effect on the earth, which is tidal in nature, is what leads to normal ocean tides, the unprecedented tidal events have to be due to objects exerting a stronger tidal force than the moon, on the earth. Since we know that Stellar Cores exert a weak gravitational force and their interaction with the Sun is tidal in nature (see Article 210: Stellar Core gravity: tidal and G is not constant) [2], the most likely objects, to be causing these tidal events, are these same Stellar Cores, which must, therefore, have attached themselves to the earth, and are drawing energy from the earth, just like they do from the Sun.

One of those tidal events, which seem to have been caused by the presence of Stellar Cores, attached to the earth, was followed the next day by a volcanic eruption, in a region just north, of where the ocean recession event occurred, suggesting a relationship between the two.

Figure 6.2. The unprecedented recession of the ocean in Kholmsk, Sakhalin Island, Russia, on March 20th, 2018 (see Article 188: What is causing the ocean to recede all over the world?) [3]

Figure 6.3. The Ebeko Vocano, on Paramushir Island, Kuril Islands, Russia, erupted on March 21st, 2018, a day after the water.

Figure 6.4. Location of Ebeko Vocano in the Kuril Islands, Russia

Besides causing water to pile up, these objects will cause severe low pressures, and thus severe storms on the earth's surface right below them. Since they seem to have now absorbed enough energy, from the earth to emit light, and since they seem to also emit bright colors of light, such as magenta (pink), burnt yellow, and orange, these colors will be observed in the sky, overhead, as they hover above, or simply, pass by, an observer on the surface of the earth.

Figure 6.5. Pink (magenta) colored sky can only occur if there is a pink light emitting light source (star) illuminating the earth's atmosphere (see Article 30: The Pink Star in our sky) [4].

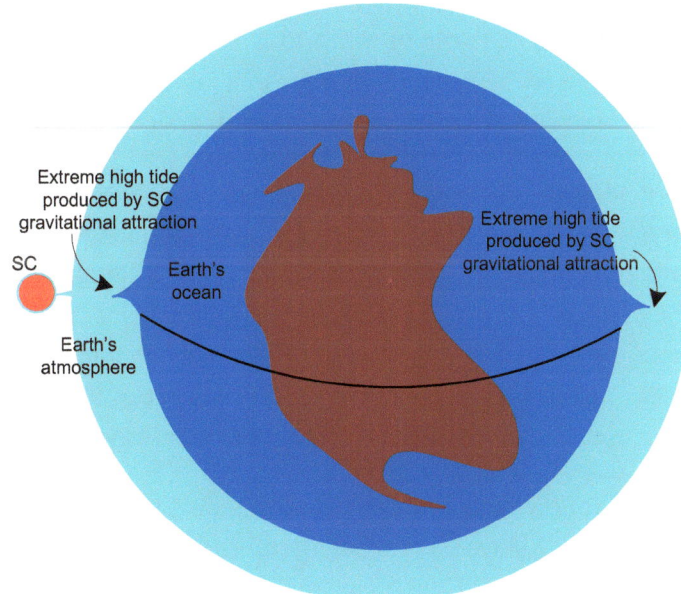

Figure 6.6. A Stellar Core (SC) close to the Earth is able to have a strong tidal effect on the earth, when it closely approaches the earth, which explains the ocean recession, observed all over the world. It will also draw on the earth's atmosphere thus causing extreme low pressure regions and thus extreme storms (see Article 188: What is causing the ocean to recede all over the world?) [3]

Now, the ocean is not the earth's only liquid layer, the earth has another liquid layer, deep inside it, and it is made of molten rock or magma. This layer forms because the earth generates a lot of energy, at its core, through fission, which causes the earth's core to heat up to a high temperature. The tidal force, or differential gravitational force due to their close proximity to the surface of the earth, will also affect this layer. The magma will be pulled toward the Stellar Core, and if the Stellar Core is in motion a tidal wave will be produced in the magma. If this wave is energetic enough a crustal displacement will result, as the earth outer solid layer disconnects from the inner solid core.

A Stellar Core, which has made earth its host will most likely be in some orbit around the earth and come in very close to the earth at times, just like they do with the Sun. The object may also at times look like a moon, and since it has become an earth satellite, we may describe it as a new moon of earth.

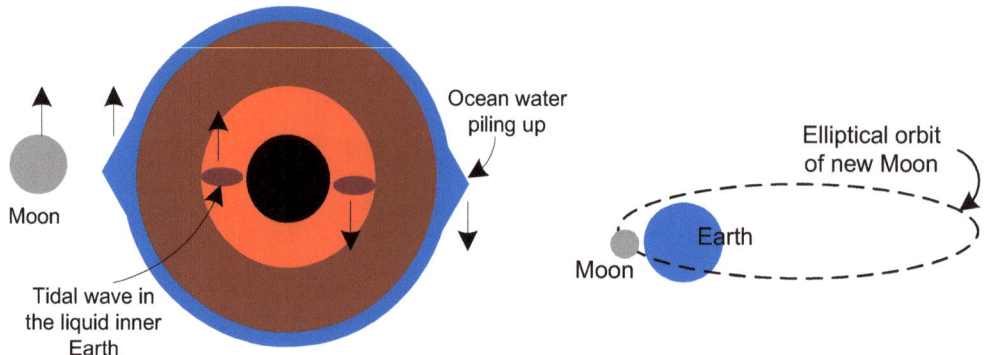

Figure 6.7. The Earth's outer solid layer can be made to shift as a result of a fast moving tidal wave generated, in the earth's inner liquid layer, by the passage of a fast moving massive object very close to the Earth. This tidal wave occurs as a result of a differential gravitational force or a tidal force.

However, these objects are far from being as benign as our moon. They are strong absorbers of the kind of energy that the earth generates, at its core, which gives rise to magma, and furthermore, as they absorb energy their gravitational influence will increase. As their gravitational influence increases their destructive effects, on the earth, will also increase, which will lead to ever stronger storm systems being created, in the atmosphere, as well more pronounced tidal effects. In addition, their gravitational pull will cause the earth to break up so that sinkholes and fissures will occur, and become more severe with time (see Article 201: Africa breaking up: a preview of what is to come) [5].

But one of their most devastating effects, on the earth, will be the fact that their pulling effect, on magma, in the earth's outer core, will cause the magma to explode, through the fissures that already exist, between this layer of the earth, and the upper mantle, and crust. This will cause the magma to push upwards, through fissures and channels, and explode through all of these all the way to the surface, wherever paths to the surface already exist. These paths will connect the magma to known volcanoes and cause these to erupt. The magma may also find new paths which would create new volcanoes.

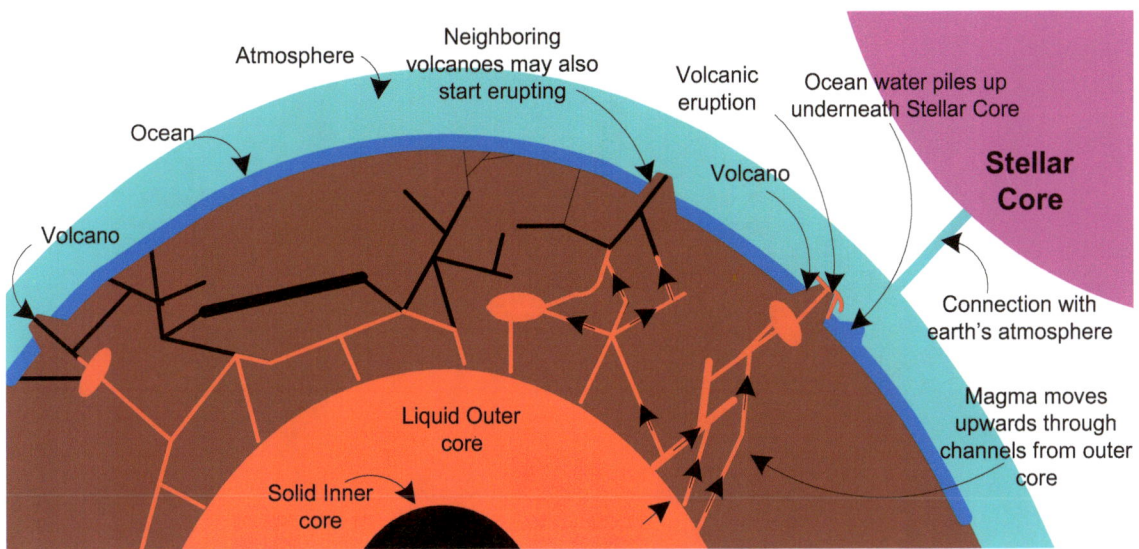

Figure 6.8. Tidal forces exerted by a Stellar Core pulls magma from the Earth's interior up through fissures and channels toward the surface, causing volcanic eruptions which are likely to increase in severity in the course of time. Neighboring volcanoes are likely to be affected and start erupting as well. This means that Yellowstone may be affected by the eruption in Hawaii.

These volcanic eruptions will become ever more violent, or longer lasting, and as the objects, will be attracted to the energy, in the molten rock, coming from the Earth's interior, these objects will be attracted to volcanic regions, where this material is closer to the earth's surface. In addition, because there are many volcanic islands, in every ocean on the planet, which are likely to erupt and continue to do so, or to do it repeatedly, to the point that part of the island, will slide into the ocean, huge tsunamis will be generated, in every ocean, and repeatedly, which will then result in all coastlines and islands, around the world, being inundated. Since most of the world's population lives along coastlines, the loss of life will be unprecedented.

In the light of the evidence from the unprecedented water recession events that have occurred in the past months it is clear that these objects are here and that it is therefore very likely that the current eruption in Hawaii which shows no sign of slowing down but rather of getting stronger as the number is related to the presence of these objects, in which case it is a sign that their stronger influence on our planet has started. This influence is not likely to weaken and this means that the volcanic eruptions will continue to increase in frequency, severity and may also be prolonged like the one in Hawaii.

Figure 6.9. Lava flows out of a fissure during the current eruption which seems to be getting more severe rather than weakening and thus likely to be a prolonged event, which may then lead to part of the island sliding into the ocean and thus create a Pacific wide mega tsunami, which would inundate coastlines.

Mega tsunamis produce waves over 100 feet in height but these can also be over 1000 feet in height. These mega tsunamis can be caused by a large meteor impact, in the ocean, or by a large amount of material sliding into the ocean. Normal tsunami's result from movement in the sea floor, usually produced by earthquakes, and produce very shallow waves out to sea, which then increases in height to about 30 feet, once it reaches a coastline and the ocean depth decreases. A mega tsunami, however, starts out with a very high wave to start with, which decreases in height with distance but can also become several times higher when the wave crosses into the shallow depths close to land. A mega tsunami occurred due to a landslide in Lituya Bay Alaska, in 1958, which produced a 1720 foot wave. During volcanic eruptions, volcanoes may undergo catastrophic flank failure, which will cause large amounts of material to fall into the ocean, at one time, thus creating a huge wave of water to flow outwards, in all directions, from that point. The volcanic Island of La Palma, in the Canary Islands, is set to create one such mega tsunami, which is likely to inundate the coast of Africa, with a 300 foot megatsunami [5].

Figure 6.11. Forecasted wave heights based on simulation of a large landslide on the Island of La Palma on the Canary Islands.

With a prolonged eruption at the Kilauea volcano in Hawaii, it also becomes a strong possibility that a landslide will occur, which will then result in a megatsunami impacting various Pacific Ocean coastlines, Simulations show that the US, Mexico and New Zealand would be impacted the most.

Figure 6.12. Wave heights based on simulation of Kilauea volcano, in Hawaii, flank failure, or landslide, showing that a wave of 100 to 150 feet, in height, may reach the west coast of the United States. Wave heights are in meters. A 50 meter wave corresponds to a 164 feet wave, which may then become twice as high, or over 300 feet, as it reaches land.

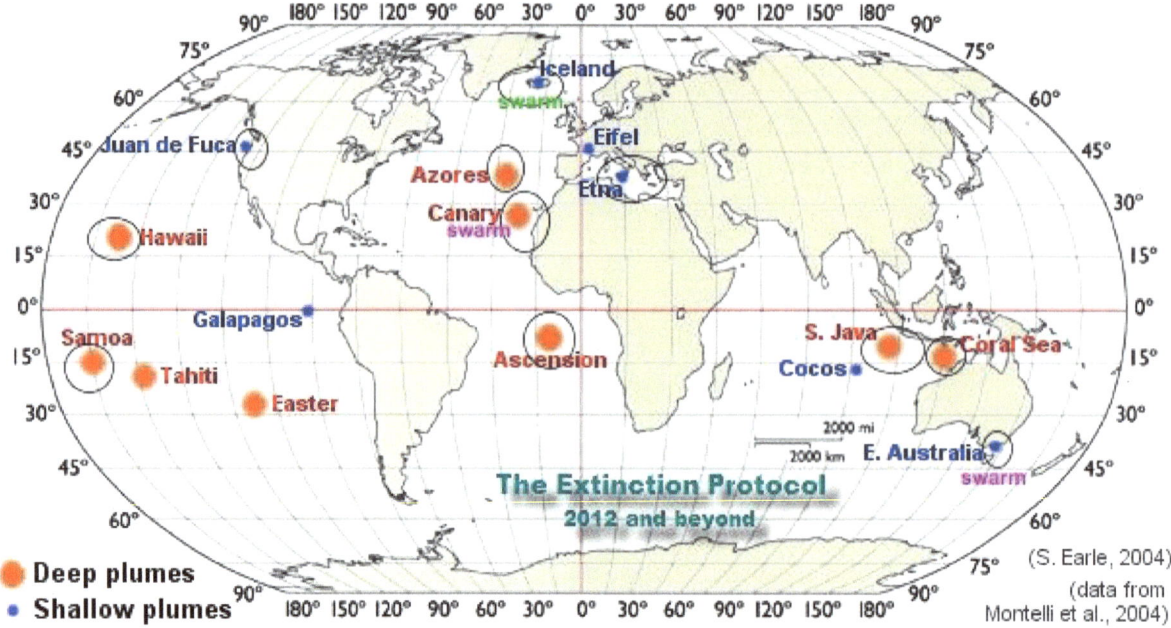

Figure 6.13. With a large number of volcanic islands around the world and the Stellar Cores' influence resulting in more frequent, more severe and prolonged eruptions, most coastlines around the world are likely to be impacted by mega tsunamis, resulting in great loss of life as most of the world's population seems to live along coastlines. This map indicates areas where the volcanic plumes are deep and thus more likely to have very large and prolonged eruptions due to the influence of the Stellar Cores which the earth is hosting. Stellar Cores are attracted to the energy produced in the center of the earth, and contained inside the magma, rising from the depths of the earth, and are thus likely to cause volcanoes to erupt over and over again as they pass overhead.

In conclusion, there is overwhelming evidence that the Solar System has been invaded by a System of dead Stars, or Stellar Cores, which are affecting the Sun, and the Earth. Clearly, some of these objects are being hosted by our planet and as their energy increases so will their gravitational effects on our planet. These effects are ocean recession, severe storms, earthquakes, fissures, sink holes and volcanic eruptions. These will continue to happen with increasing strength and severity. Severe and prolonged volcanic eruptions, such as the one now occurring in Hawaii, are likely to also lead to landslides and mega tsunamis, and since volcanic islands occur in every ocean, on the planet, all coastlines are in danger of being affected and thus inundated by mega tsunamis.

References:

[1]   Albers, C. (2018). Article 236: What are Stellar Cores?
[2]   Albers, C. (2018). Article 210: Stellar Core gravity: tidal and G is not constant.
[3]   Albers, C. (2018). Article 188: What is causing the ocean to recede all over the world?
[4]   Albers, C. (2017). Article 30: The Pink Star in our sky.
[5]   Albers, C. (2018). Article 201: Africa breaking up: a preview of what is to come.
[6]   http://www.geotimes.org/oct04/NN_tsunami.html

Chapter 7

238: Aliens exist

Do aliens exist? I would say that the evidence is quite overwhelming that they do exist. But let us begin with a definition of what the word 'alien' means. The word 'alien' refers to something not of this earth. This definition encompasses living beings, organisms, and technology; anything that does not originate on earth. The evidence available regarding the question of whether aliens exist, comes from both what many people have witnessed, the many videos and photographs showing craft in the sky, doing maneuvers that are impossible, for any normal craft, and is also available from the many ancient monuments, which have been built with technology that is not yet available to us.

Figure 7.1. The pyramids of Egypt have been built in such a precise manner that they could not have been built through any of the modern methods available to us now. This means that they have been built with alien technology, by either alien, or by humans, with access to that technology. Either way, that is evidence that aliens exist.

Pyramids exist all over the world, and at the bottom of the ocean. A blue crystal pyramid has been discovered, at the bottom of the ocean, in the Bahamas [1], in a region long known to be strange, as unexplained phenomena have been occurring in this region, for a very long time.

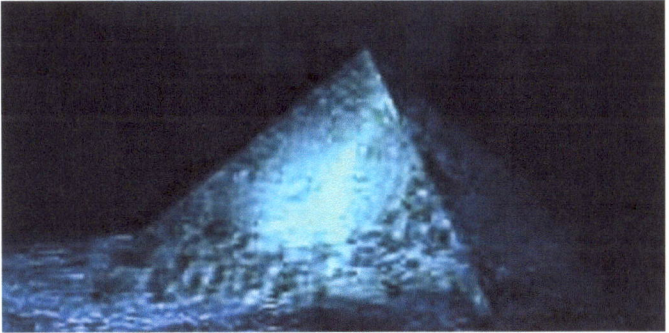

Figure 7.2. A blue crystal pyramid has been discovered, at the bottom of the ocean, in the Bermuda Triangle [1].

Figure 7.3. Left: A pyramid in Mexico. Right: A pyramid in China.

Figure 7.4. Another pyramid in China, this one like many others, is covered in soil and vegetation, but the overall shape is unmistakable. Natural mountains do not form in the perfect shape of a pyramid.

Figure 7.5. Ancient pyramid complex, discovered in Bosnia.

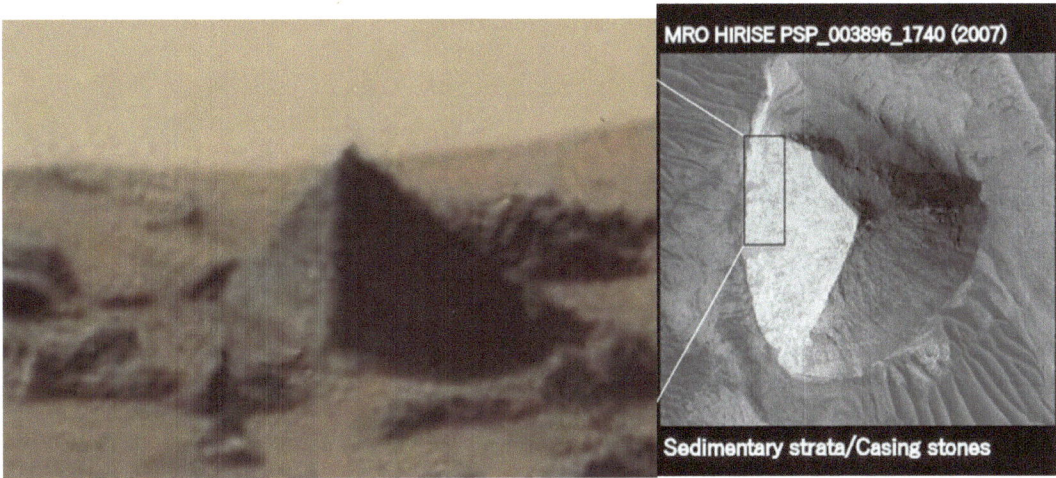

Figure 7.6. Left: A pyramid found on an image of the Martian surface [2] **Right:** 3 sided artificial structure found on Mars [3].

Now, if pyramids have clearly been built with technology that is not available, at the present time, and if they exist on another planet, is that not clear evidence that aliens exist? If they were built on another planet, than that is in itself evidence that intelligent life capable of developing technology exists, and the fact that these same structures appear here, and we could not have built them, is evidence that this same intelligent life, which is alien to our planet, was here in ancient times and built these structures. Or, as I said before, they passed the technology they had, to certain human beings, who then used it to build these structures.

Is there also evidence that aliens are doing the same thing nowadays, passing their technology on to certain human beings, who are then using it, and hiding it from the rest of the population, on this planet? I would say that there is also evidence of that. There is advanced technology is being used on this planet right now; advanced craft, with cloaking capabilities, are being used in our skies, in order to spread chemtrails [4]. And, artificial holographic technology is being used in our skies, to produce an artificial sun, in the atmosphere [5].

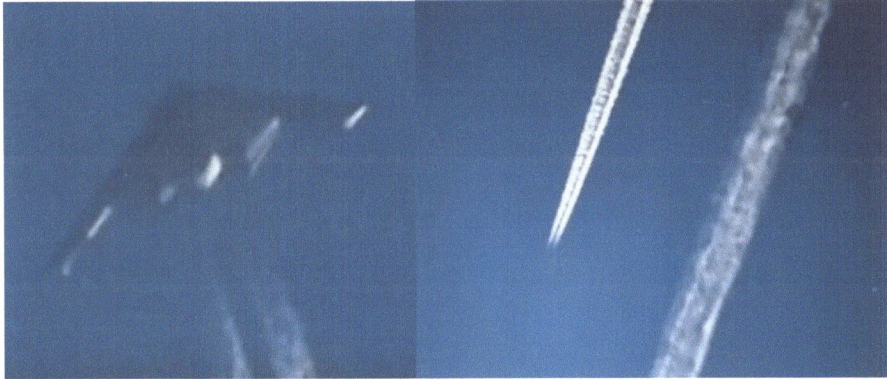

Figure 7.7. Left: A stealth aircraft spray aerosols. **Right:** No aircraft can be seen in front of the double white line indicating that the aircraft responsible is cloaked.

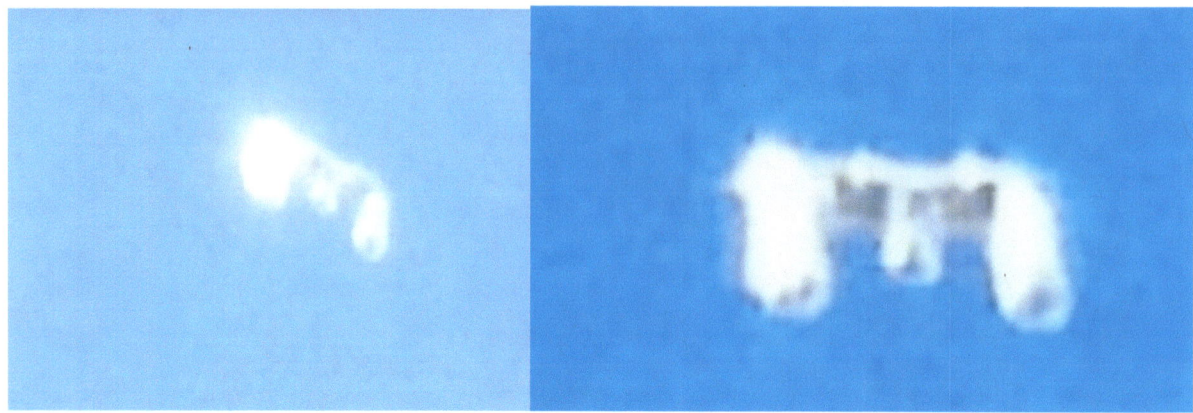

Figure 7.8. Artificial device with what seems like projectors, and screens, on it, suggestive of what a holographic projecting device may look like.

Figure 7.9. Screenshots of footage, taken from an airplane, showing the Sun simulator in the clouds: This footage was shown in a Youtube video by Shahzwar Bugti. The bright yellow part seems to keep appearing besides the airplane, even as it moves forward, indicating that a lens system is in operation, in addition to, a lighting system. It seems that the cloud is being used as a three dimensional screen, for a holographic type of projection. The bright yellow part also appears to be quite small from the airplane's perspective.

There is also evidence that these aliens passed some of their technology onto the NAZIs, in the 1930s, and that they have passed on their technology to the US military, and possibly to the military of other countries. When a country goes from flying biplanes to building antigravity drives, and zero-point energy power systems, within a few short years, where is the logical place they got their sudden phenomenal breakthroughs from? We know how technology invented by human beings leads to progressive step by step advances. A sudden advance of this magnitude indicates that the technology was not invented by human beings, but by beings who had advanced their technology to a much higher level than humans and then passed that knowledge on to human beings.

But where do these beings come from, and what are their intensions, should we blindly trust them, and think that they are benign, or should we seek to understand who, and what, they are, and perhaps, as well, what they want with us? I think it is very naïve to just trust them. One of the reasons why we should not trust them is that the technology they give away is usually given away to elite groups only, who then not only hide it from the general population; they also use it against the general population. Think about it, computers, internet, cell phone technology is often seen as one of the greatest advances of our civilization. On the other hand, the technology is clearly being used to spy on and control us, and cell phone towers emit radiation that is harming us and causing disease. There are certain whistle-blowers, such as Robert Lazar, which have come out and said that this same technology came from a downed alien craft, which was reverse engineered, in places like Area 51 [6].

The question needs to be asked then. If their technology is so advanced, why are their craft crashing so easily? Is it possible that they are crashing on purpose so that our military would be able to get their hands on the technology? And since the technology seems to be used to enslave, rather than to free mankind, how can we say that this is a good thing? Or that their intensions are good?

Figure 7.10. Clouds in the sky, which look like saucer shaped craft, and thus evidence of the existence of alien technology, on earth: It is not natural for 5 clouds to look so exactly the same and to have such a symmetrical shape. These have to be cloaked saucer shaped craft. This craft clearly have cloaking capabilities, which makes them look like a cloud.

So we have evidence that aliens exist and that they have been around for a long time since we know they were around when the pyramids were built, and most likely before then, since pyramids even older than the Egyptian pyramids, have been discovered all over the world, and on Mars. Where else can we go to investigate who these beings are, and what their intensions are? There are many ancient writings in the world, which talk about these beings. Most of these writings say the same basic things about them. They seem to like creating an elite, in whatever culture they get into contact with, which then enslaves the common people.

Figure 7.11. With the alien technology given to them, the elite get to live in a type of paradise, in space, whilst we on the ground are tortured by mind control weapons and purposely given cancer, whilst being denied the treatments that easily cure it, and are available to the elite through their alien masters

There is one ancient book, which seems to have a different agenda, though, and this is the Bible. This book tells the story of an enslaved people that were freed by their God. This God did not create an elite, which then proceeded to enslave their fellow man. No, this God gave them laws, which pertained to freeing those who had fallen on hard times every 7, or 50 years. This God gave them such laws such as: do not harvest everything out of the trees, leave it, this is to provide for the poor of the land, and the strangers:

[22] *"'When you reap the harvest of your land, do not reap to the very edges of your field or gather the gleanings of your harvest. Leave them for the poor and for the foreigner residing among you. I am the LORD your God.'"*

Lev. 23:22

Figure 7.11. A very large disk shaped cloud and most likely a cloaked ship. The circular funnel like cloud formation indicates that the atmosphere is more negatively charged just below the craft, thus leading to cloud formation underneath it.

The being that calls himself 'the LORD your God' is not of this Earth, and is, therefore, an alien according to our definition. This God, though, seems to care for human beings, so maybe we should listen to what he has to say about other aliens. Does the Bible mention aliens? Yes, they are mentioned in two places, in Genesis. We are first of all told the story of the serpent who lied to, and beguiled, Eve and caused Adam and Eve, to lose their place and connection with God. The story is woven then, through the Bible of how God changed his name to Satan, which means enemy, and God said in Genesis that one would come who would destroy the serpent. Now, if a being purposely lies in order to cause harm to our forefathers, Adam and Eve, would we not say that he had bad intentions? I would think so, yes. Does the Bible identify this being further? In Isaiah 14:12-14 it says:

12 How art thou fell from heaven, O Lucifer, son of the morning! how art thou cut down to the ground, which didst weakens the nations!

13 For thou hast said in thine heart, I will ascend into heaven, I will exalt my throne above the stars of God: I will sit also upon the mount of the congregation, in the sides of the north:

14 I will ascend above the heights of the clouds; I will be like the highest.

So we are told that his name is Lucifer, he fell from heaven, that he was able to weaken the nations of the earth, and that he had a goal of raising his throne, which means that he was a ruler, to the height of the God of the universe, the highest. This same ruler tempted Jesus, and Jesus placed himself in opposition to him, as a thief in John 10:10:

[10] The thief comes only to steal and kill and destroy; I have come that they may have life, and have it to the full.

And as a murderer and a liar, in John 8:44:

Ye are of your father the devil, and the lusts of your father ye will do. He was a murderer from the beginning, and abode not in the truth, because there is no truth in him. When he speaketh a lie, he speaketh of his own, for he is a liar and the father of it.

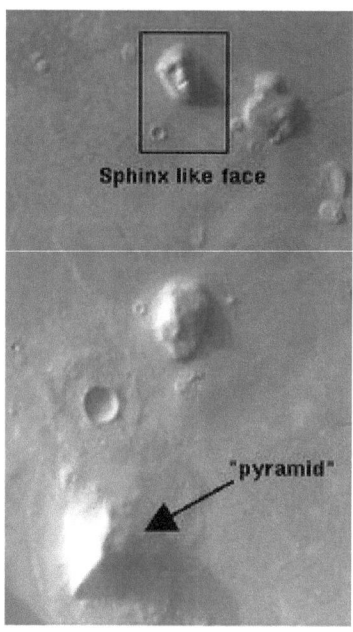

Figure 7.12. Signs of an intelligent advanced civilization on Mars and that they have a face with a mouth, and two eyes, like us.

The story we are told in the Bible is that the alien being, who came into the garden, lied to Eve, and through his lying took control of the planet, and the nations on earth, he took the human race captive at that point. Jesus came to free us from the slavery that was enforced on the human race. He sought to give life but yet Jesus was rejected over and over again, why? Because this alien ruler has a way to talk to every human being on earth, and get them to do his bidding, and every human being is born in that state, because of what happened in the Garden of Eden. Freedom comes only through acknowledging God as Lord, if this is not done, Lucifer remains the Lord of that person. This being tempted human beings, even those who are followers of God to do evil. And yet when you listen to people who have had contact with aliens, they usually have the concept that we human being are still too spiritually immature to know the truth or join the intergalactic community. Why do they never say that human being are purposely, telepathically, and continuously, invited and pushed, to do evil to themselves, and their fellow human beings, and that this is the main reason why there is so much trouble in the world?

Figure 7.13. A large disk shaped craft appears in the sky, evidence of alien technology on earth.

Why do they not ever acknowledge that the nutrition has been purposely taken out of food, which is known to cause violence and anger in human beings, and that poverty is purposely created at the same time so that there is a crime? Who is directly responsible for these horrible crimes against humanity? Is it the rich, and powerful elite, or the one who made them rich and powerful? Yes, it is Lucifer, the chief alien, on this planet, who is responsible. Why do the aliens who contact human beings never mention this? Because Lucifer is their leader.

Are there good and bad aliens? Yes, there are those who have Lucifer as their leader and there are those who have God, the highest, the creator of the universe, and creator of every angelic or spiritual being, including Lucifer. The aliens who come from God will tell you they come from God, and acknowledge the truth regarding Jesus. They will say that Jesus came in the flesh, and the Bible teaches us to check this, by asking them to say that Jesus came in the flesh. We need to check because Lucifer is a liar, and all those under him, will lie as well.

So where do all the alien races come from? There are 2 types of aliens, physical beings, who can die, and aliens that are spiritual beings, and cannot die. The physical beings have been created, through genetic manipulation, Lucifer seems to be good at genetically creating new races by mixing human DNA, with the DNA of angelic, or spiritual, beings like him. This started out with a few of these angelic beings that came down and took human women as wives, before the flood as reported in Genesis. The woman then gave birth to giants, who were only half human. Since then Lucifer has, through genetic tinkering, created several races: Blue Avians, Nordics, Greys, etc. which then pretend to have originated on a planet far away. They may have their home base on other planets, and even other star systems, but they are here because they are part of the grand deception that Lucifer has woven. He was a liar from the beginning, and there is no limit to his deceptive ways.

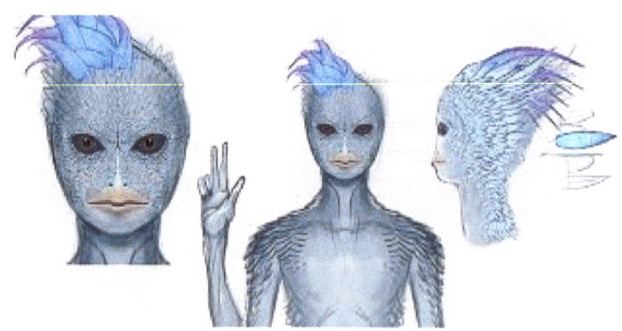

Figure 7.14. Alien race: Blue Avians.

It is usually easy to see that these beings are actually inferior to human beings, by the fact that they are delicate, and easily killed by humans, who have the greatly inferior technology. Furthermore, they are infertile, which is a sign that they are created through genetic manipulation, rather than being directly created by God. It is well known that these beings reproduce through cloning and not directly as we do. It is also a well understood concept in agriculture that weak plants are not able to reproduce normally, but have to be planted each time, or cloned, and that these weak plants are often hybrids, or genetically manipulated. Clones are exact copies. Uniqueness in every individual, of a species, is God's trademark, Lucifer's technology is far inferior to that.

So what are we to do? We are now living at a time when tyranny is rising up, and the rights of normal people are being taken away, at every corner, we are being bombarded by radiation, from cell phone towers, chemtrailed as if we were an insect pest on the earth, and having poisonous chemicals in our water, as well as, having poisonous genetically engineered food forced upon us. Do we rise up and try to take power away from the elites that seem to be oppressing and killing us? In the light of the understanding of who is really ruling this planet, this will do us no good. Lucifer will just form another set of elite and continue to oppress us through them. We have to realize that the problems we face are spiritual in nature because Lucifer is a spirit being. We need to turn to our creator, get on our knees and make him the Lord of our lives. If every human being on earth did this all the problems we face would be resolved. But in the absence of that, we need to take this step as individuals.

In addition, we should realize that lying and steal places us in Lucifer's camp because that is his way of operating, so if we have ever stolen or lied, we should repent, which means change and do the opposite from now on, ask God to forgive us, and stop doing what the enemy of God, and mankind, does.

In conclusion, aliens exist; the chief amongst them is the ruler of this planet, Lucifer. He has created various races, who answer to him. He has the very advanced technology available to him. He is also the source of all misery and tyranny on this planet. We need to clearly see who the enemy is, in order to know who we can, and cannot trust. It is not wise to trust those who lie in order to enslave us. It is wise to trust those who tell us the truth and give their life to free us. We should, therefore, trust the real God, and creator of the universe, who has proven his love for us and stop being taken in by those who like to pretend that they are God.

References:

[1] https://www.disclose.tv/atlantis-and-the-blue-crystal-pyramid-in-the-bermuda-triangle-310025

[2] http://home.bt.com/news/science-news/pyramid-found-by-nasa-curiosity-rover-is-proof-of-intelligent-life-on-mars-say-ufo-spotters-11363988251648

[3] https://www.ancient-code.com/artificial-structure-mars-three-sided-pyramid-verified-science-journal/

[4] Albers, C. (2018). Article 231: Advanced technology in the sky.

[5] Albers, C. (2018). Article 233: the holographic Sun in the clouds.

[6] https://science.howstuffworks.com/space/aliens-ufos/area-517.htm

Chapter 8

239. Primary and secondary alien technology

People are usually in awe of the type of technology that aliens, who use saucer shaped craft to get around, have. In this article, I would like to show that their technology is actually much like human technology and very inferior to what the creator of the universe is capable of creating. As detailed in Article 238: Aliens exit [1], an alien is something or someone, that is not of this earth, but there are two distinct types of aliens, aliens who have aligned themselves to Lucifer, and aliens that are aligned to God. God, the creator of the universe, is not of this earth and is thus an alien too. There are two main types of technology, in the universe, God, or creator, technology, which encompasses all that naturally exists: galaxies, stars, planets, energy or light and particles. It also includes living organisms, such as we find on planet earth, which procreate through various natural means, for examples, plants produce seeds, which then grow into a plant of the same species. The daughter plant will have characteristics which identify it as belonging to the same species, but it will also be different from its parent, it will be a unique individual, within the species. We will call all that type of technology, primary technology. The other type of technology is the kind that takes that which naturally exists and manipulates it, in order to create something out of it, or use it to perform a function, such as to transform it into a form of energy, which can then be used to run an engine, or a whole city. It can also refer to the manipulation of naturally occurring plant and animal life, through the use of genetic manipulation to create different forms of such life. We will call this type of technology, secondary technology.

Figure 8.1. Left: Disc shaped craft with antigravity driving technology. Right: A space station in orbit, which may be built by humans, using reverse engineered secondary alien technology. Secondary alien technology, like technology invented by human beings, is used to produce artificial devices. See Article 238: Aliens exist [1] for the details on aliens and how we can know that they exist.

Now, the alien races that are visiting earth and that do not claim to be coming from the true God, seem to have very advanced driving systems for their ships, but their technology is still of the secondary type.

They may produce ships that are the size of moons, and fill them, with perhaps, 100 000 individuals, but all of it will be produced out of materials that already existed, and thus they use primary technology to produce secondary technology. The individuals themselves will procreate through cloning and thus through a secondary, not a primary process. They will all be exact copies of their parent. That is another facet of secondary technology, it is not easy to produce a functioning device, and once one is produced, it can be copied but each individual copy is an exact copy of the original. In the case of the primary type of technology, however, every individual is unique and often different parts, of each individual, are unique as well.

Figure 8.2. A hollow moon, with a city inside it, is an example of secondary technology. It will contain transport systems, manufacturing, and power generation facilities, as well as storage facilities and waste disposal systems, and all may be coordinated by a central computing system; all built out of naturally existing materials.

For instance, a small plant may have 10 000 leaves on it, and each is different from the other, the main components of each leaf are cells, each cell has a diameter of about 100 micrometers. But each cell is a living entity with a nucleus, which is like a central processing center, controlling all the functions of the cell, such as growth, metabolism, protein synthesis and reproduction, or cell division. Proteins are the materials needed to build structures and thus each cell has a manufacturing facility. There are also power plants inside a cell, called chloroplasts, which convert light into energy, through a process called photosynthesis; there are also storage facilities, in cells, called vacuoles, where nutrients can be stored. These vacuoles will, however, double up as garbage disposal facilities, as they contain enzymes that can break up unwanted molecules, in a cell. Thus, a cell seems to have everything that a whole city requires. A central headquarters or processing center, a power plant, storage facilities and garbage disposal, all within something with the diameter of 100 micrometers.

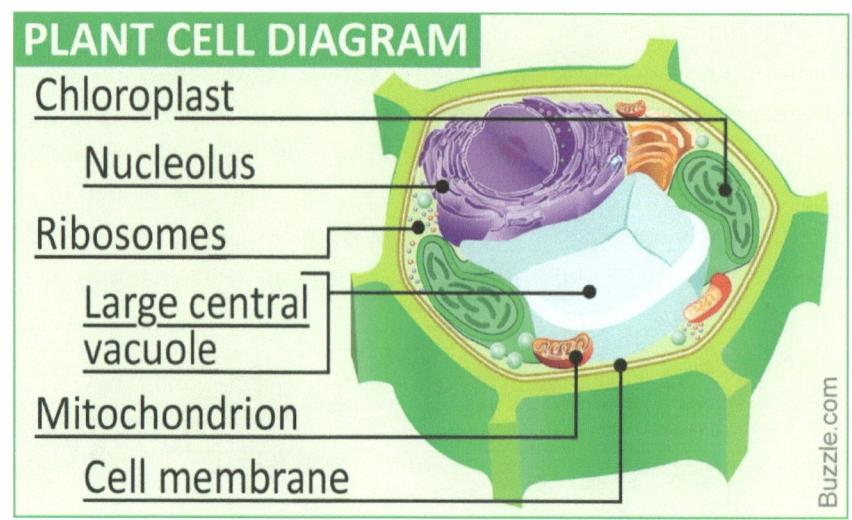

Figure 8.3. Illustration of a plant cell: the nucleus is like a central processing and material manufacturing unit, the chloroplast is like a power plant converting light to energy the cell can use, the vacuole functions as both a storage and a garbage, disposal unit, all within a width of 10 to 100 micrometers. A human hair varies between 20 and 200 micrometers, in thickness, so these cells are usually only half the size of a human hair. They also have transport systems, within them, so that nutrients can get to the right places and waste can be removed.

Figure 8.3. A leaf of 2 by 5 cm, in size, may contain 100 000 cells, each cell works like a miniature city. Water and nutrients flow in, individual proteins and enzymes are manufactured and everything works together, precisely and on time.

So how many of these primary technology functioning cities are found in one leaf, on a plant? Let us suppose we have a leaf which is 2 by 5 cm in surface area, this would be about 1 by 2 inches. Plant cells actually vary in size between 10 and 100 micrometers in size; we are going to assume that each cell occupies an area, with a radius of 100 micrometers, to make sure we account for the area between cells. We will also estimate that there are 3 levels of cells, in each leaf. Then, the surface area of one leaf is 10 cm², or: A = 10^{-3} m². So, the numbers of cells in one leaf is:

$$N_{cells} = 3\frac{A}{\pi r_{cell}^2} = 3\frac{10^{-3} \text{ m}^2}{\pi(10^{-4} \text{ m})^2} = 9\times10^4 = 100\,000 \text{ cells}$$

where the 3 accounts for the number of levels of cells, in each leaf, and r_{cell} is the radius of each cell. We thus obtain 100 000 cells or complete miniature cities, in each leaf, with a plant containing some 10 000 leaves, we then end up with 1 000 000 000 or 1 trillion mini cities, and on top of that, each individual leaf is unique. Now, if we then consider the number of plants on the planet, which belong to the same species of plant, we can see that we quickly get to incomprehensible numbers. This should, however, give an idea of the difference between primary and secondary technology and therefore the difference between that which Lucifer and his subjects are capable of creating, and what God is capable of. There is just no comparison.

In conclusion, Lucifer's technology is a secondary type of technology, and far inferior to the primary technology of the creator of the universe, the true God.

References:

[1]   Albers, C. (2018). Article 238: Aliens exist.

Chapter 9

240. Planet X System effect on radioactive decay rate and heating of planets

I use the term Planet X to refer to the system of dead stars, or Stellar Cores, which have been extensively observed in the Sun's corona. These objects are no longer able to generate enough energy in their core to maintain a negative outer layer and to emit light. Their low energy status also results in a very low strength gravitational interaction between them and other objects.

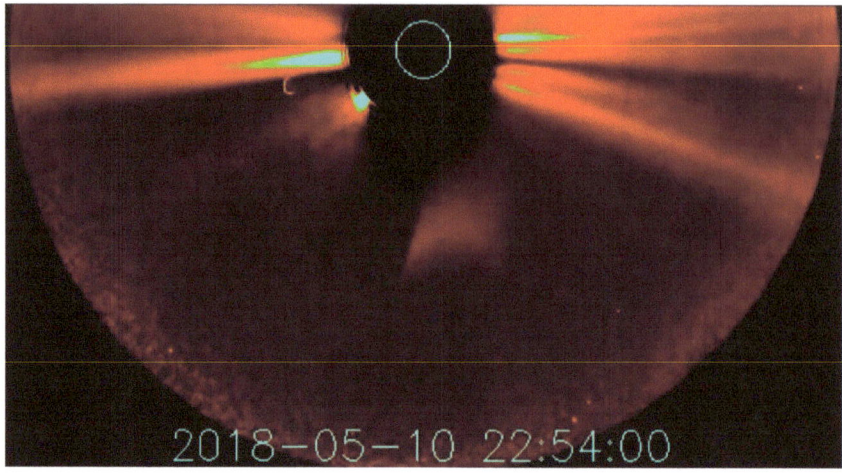

Figure 9.1. A Stellar Core in the Sun's outer corona. The object seems to have a trail, behind it, of material it is either drawing from the Sun's corona or leaving behind it. These objects are known to shed their outer layers, when in the Sun's corona. The trail is partially obscured by the black cloud underneath the Sun and must, therefore, originate from under the Sun.

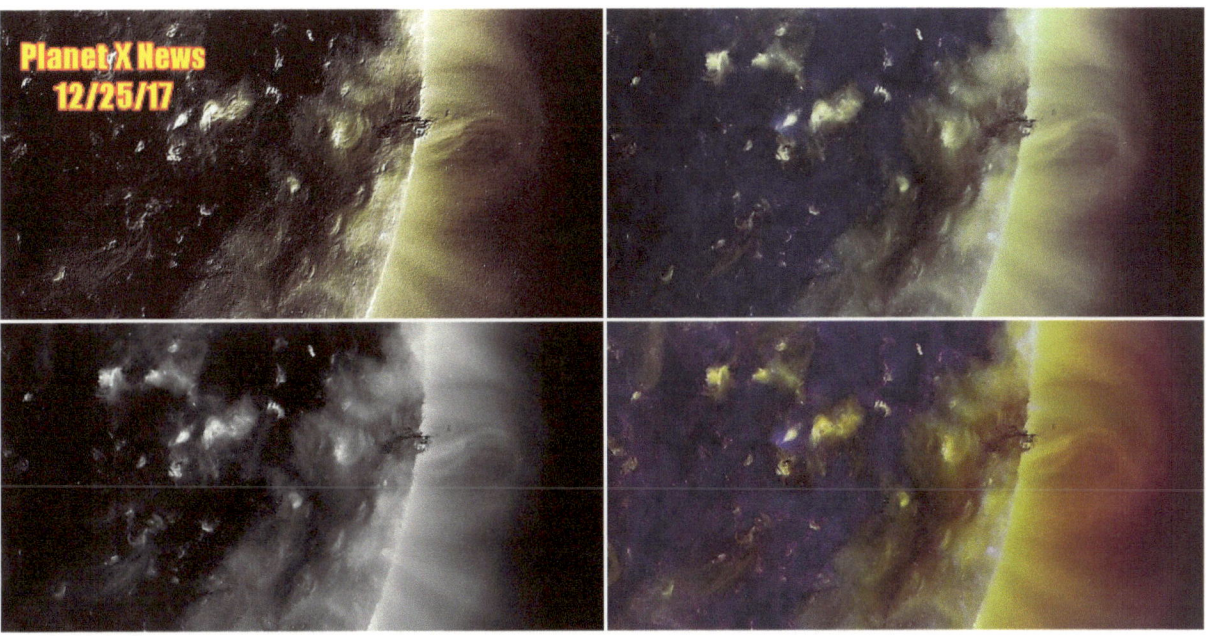

Figure 9.2. A Planet X Object or Stellar Core appears in the Sun's corona. The object is striped and a size comparison, with the Sun, reveals that it is about 4 times larger than the earth. It has a similar trail to the object in figure 1 behind it.

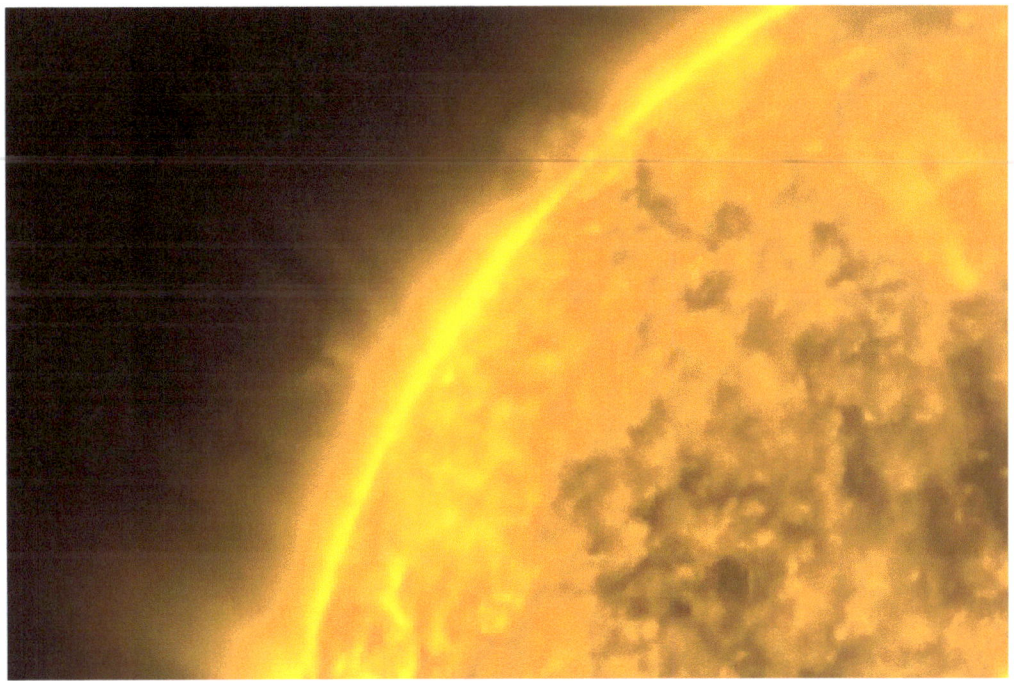

Figure 9.3: SDO image in the 171 angstrom wavelength from October 13th, 2017 showing a dark Stellar Core, which appears to be about half of the size of Jupiter. The object is making a matter of connection with the Sun, through which it is drawing energy from the Sun. The objects draw energy through direct absorption of light photons, as the 2007 Stellar Core demonstrated, when it traversed the Sun (see

Article 183: Stellar Cores absorb photons) [1], and also draw energy through drawing matter, as this matter contains gravitational energy, in the form of photons (see Article 184: Stellar Core evolution) [2].

The low strength of the gravitational interaction, between the Stellar Cores and Solar System objects, requires that the gravitational constant, G, for any gravitational interaction between these objects and other celestial objects be very low, i.e. less than 0.6% of normal, in the case of the 2007 Stellar Core, which traversed the Sun in February of 2007 [3]. Thus, we could say that

$$G_{SC} \leq 0.006G$$

where $G = 6.7 \times 10^{-11}$ N.m^2.kg^{-2} is the strength of the gravitational constant between Solar System objects. In Article 153: Planet X: Escape velocity and Gravity [4], I calculated the reduced gravitational effect of these objects in terms of an effective mass, but since then I have understood that mass is not affected, it is the strength of the gravitational interaction, which is affected. In other words, it is the gravitational strength G, in the gravitational interaction equations, that is affected, and not the mass of the objects. The equation for the force, between any two masses, will thus still be given by:

$$F_G = G \frac{Mm}{r^2}$$

But the G can no longer be referred to as a constant, as it is actually a function of the object's energy and will thus will vary according to the energy status of the object, i.e. $G = G(E)$. Now, since the energy generated by a star, or planet, in its core, is directly related to the strength of the gravitational interaction and thus the gravitational field it generates, which in turn is responsible for the electric field it produces, and since this energy is generated by radioactive decay, it is likely that the gravitational interaction affects radioactive decay rates, which will thus not be constant, either. If the radioactive decay rate speeds up, as G decreases, this provides yet another process through, which Stellar Cores can affect the Earth, and indeed, all objects in the Solar System, including the Sun.

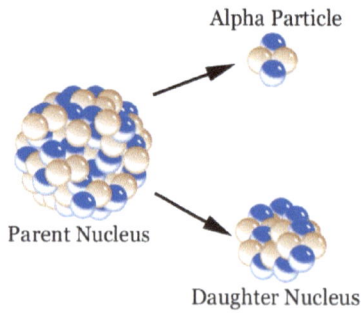

Figure 9.4. Radioactive decay is the breaking up of a heavy nucleus into 2 lighter nuclei. A drop in the gravitational energy of the particles in the nucleus will decrease the attraction between protons and allow the electrostatic repulsion between them to pull the nucleus apart.

In the light of how the gravitational interaction works, and how the strong force, which works inside the nucleus to keep it together, is actually attributed to the gravitational interaction [5], it becomes clear that the decay, or break up, of a nucleus has to be related to the strength of the gravitational interaction. If the strength of the gravitational interaction drops then the attraction between protons drops, which allows the repulsive electrostatic force, between protons, to force the nucleus apart, and thus to decay. Since the strength of the gravitational interaction is dependent on the gravitational energy, a drop in gravitational energy is likely to lead to the break up to happen more easily, and therefore, for radioactive decay to occur at a faster rate.

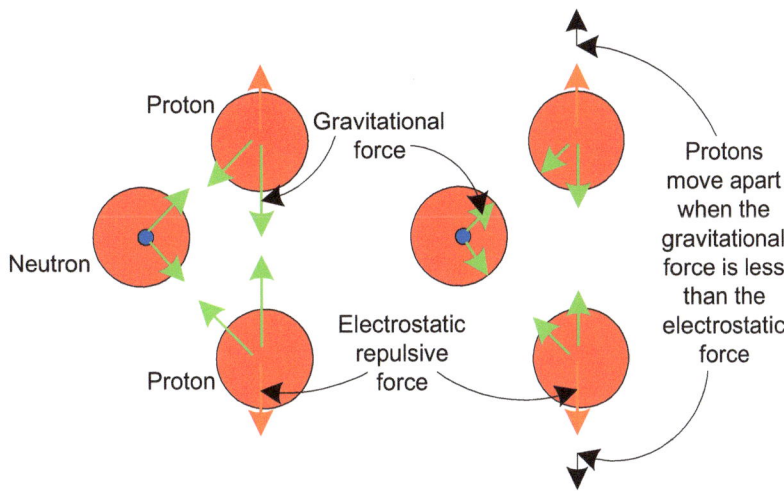

Figure 9.5. When the gravitational energy drops the gravitational attraction between protons also drops. If the electrostatic repulsion is stronger than the gravitational attraction, the protons will tend to move apart, which causes the nucleus to break into parts

Gravitational energy generated in the core of a celestial object (star or planet) will most likely flow from particle to particle, toward the surface, and may be thought of in terms of heat, but are actually light photons. Particles in close proximity to each other will share the energy that is available, so that any two particles, in close proximity, will end up with the same amount of energy. Thus, the energy flows outwards, from the core, toward the outer layers, and the rate at which it flows will reach a point, which is constant. This means that the flow of energy is in a state of equilibrium, called the steady state, in which the rate of energy generation is in balance, with the heat emanating from the surface. At this balanced state, the surface remains at a constant but much lower energy than the interior or core, which in most cases results in only infrared radiation, being emitted from the surface. This is the same for both established planets and stars. Venus is not in this state yet, as it seems to be a newly formed planet (see Article 169: Planetary formation: comets to planets) [6]. The Sun has to have a cool surface, as it goes completely dark, in response to the presence of certain Stellar Cores (see Article 232: The Sun can go dark: the implications) [7]. The Sun's light emission and high atmospheric temperature arise from ionization, of its atmosphere, as a result of the electric field generated in its atmosphere, which is in turn due to its gravitational field, or gravitational energy, generated in the core.

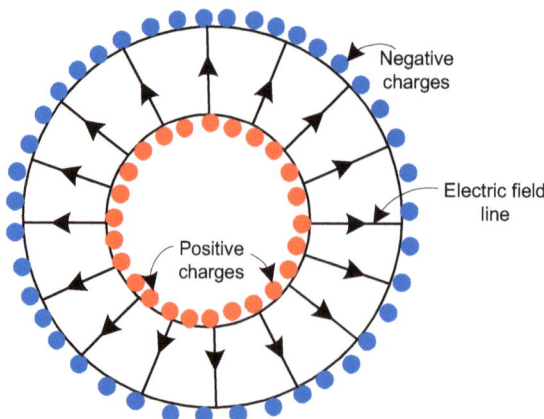

Figure 9.6. The gravitational energy generated in a star's or planet's core gives rise to a gravitational field, which causes protons, to attract, and protons and electrons, to repel, and thus to protons and electrons to separate. When positive and negative charges are separated, then an electric field is generated, in the space separating the two. Thus, an electric field is generated between the Sun's core and its outer negative layer and therefore in its atmosphere.

But the Planet X System, of Stellar Cores, absorbs gravitational energy from the Sun and planets, in the Solar System, which will tend to pull these out of a state of equilibrium, as it will cause the gravitational energy, from the core, to flow outwards at an increased rate, and thus to drop in the core. This will trigger an increased rate, of radioactive decay, and thus, of energy generation, in the core. Thus, the planet maintains its core gravitational energy, but more heat flows to the surface, which thus heats up. This then offers an explanation as to why all the planets, in the Solar System, seem to have been warming up, for many years [8], and why earth seems to be warming up from the interior outwards. Increased heat will lead to more volcanic eruptions, and since most volcanoes are at the bottom of the ocean, this will cause earth's oceans to greatly warm up [9].

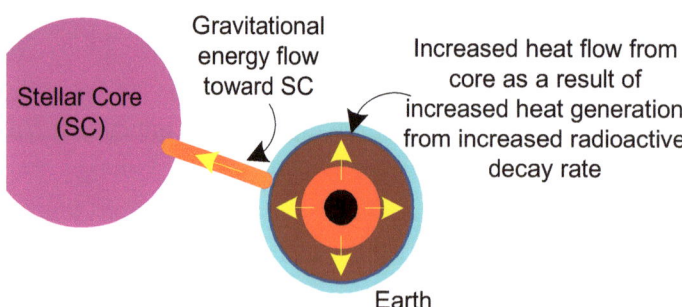

Figure 9.7. Stellar Core's absorption of gravitational energy (photons or heat), from a planet, decreases the gravitational energy in the core, which causes the rate of radioactive decay to increase and thus the heat generation to increase, which increases the rate of heat flow to the surface and thus increases the planet's surface temperature. In the case of the earth, it has greatly increased its ocean temperature.

Now, the rate at which gravitational energy is generated in the core, of a celestial object, will affect the whole object and its overall gravitational field, so the radioactive decay rate will change everywhere on the celestial object, not just in the core.

The evidence that Stellar Cores are being hosted, by the Earth, is overwhelming and lies in the many ocean recession events that have occurred all over the world. Only a tidal force of a greater magnitude than that, which the moon exerts, on the Earth, is able to account for such events. For more details, see Article 188: What is causing the ocean to recede all over the world? [10] and Article 227: Stellar Cores affecting the earth and possible connection to volcanic eruptions [11], as well as, Article 237: Hawaiian Island and Mega tsunamis [12].

Figure 9.8. An empty beach, due to the ocean receding, from the Brazilian coast, on August 12[th,] 2017, no large storms or hurricane could be blamed for the phenomenon, as there were no storms or hurricanes anyway near this coastline. This event was unprecedented.

In conclusion, radioactive decay just like the gravitational constant is dependent on the gravitational energy status, of a celestial object, and it is thus not constant. Stellar Cores absorb gravitational energy from celestial bodies, in the solar system. This results in increased radioactive decay rate, which increases the rate of energy generation, in the core, and also in an increased surface temperature. Increased radioactive decay rate as a result of the presence of the Planet X System, therefore, offers an explanation as to why all the planets in the solar system seem to have warmed up, in the past decades [8].

References:

[1] Albers, C. (2018). Article 183: Stellar Cores absorb photons.
[2] Albers, C. (2018). Article 184: Stellar Core evolution.
[3] Albers, C. (2018). Article 143: Planet X traverses the Sun: irrefutable evidence
[4] Albers, C. (2018). Article 153: Planet X: Escape velocity and Gravity.
[5] Albers, C. (2018) Book: Planet X Revealed Gravity and Light. Amazon Publishing.
[6] Albers, C. (2018). Article 169: Planetary formation: comets to planets.
[7] Albers, C. (2018). Article 232: The Sun can go dark: the implications.
[8] https://www.livescience.com/1349-sun-blamed-warming-earth-worlds.html
[9] https://insideclimatenews.org/news/03102017/infographic-ocean-heat-powerful-climate-change-evidence-global-warming
[10] Albers, C. (2018). Article 188: What is causing the ocean to recede all over the world?
[11] Albers, C. (2018). Article 227: Stellar Cores affecting the earth and possible connection to volcanic eruptions.
[12] Albers, C. (2018). Article 237: Hawaiian Island and Mega tsunamis.

Chapter 10

241: The Rapture at Pentecost

In this article, I would like to detail something which I have observed in the Bible regarding the timing of the rapture and how it may be connected with Pentecost or Shavuot (the Feast of Weeks). First of all, I would like to define what I mean by the word 'rapture'. I use this word to refer to the event described in I Thessalonians 4:13 to 17:

³ Brothers and sisters, we do not want you to be uninformed about those who sleep in death so that you do not grieve like the rest of mankind, who have no hope. ¹⁴ For we believe that Jesus died and rose again, and so we believe that God will bring with Jesus those who have fallen asleep in him. ¹⁵ According to the Lord's word, we tell you that we who are still alive, who are left until the coming of the Lord, will certainly not precede those who have fallen asleep. ¹⁶ For the Lord, himself will come down from heaven, with a loud command, with the voice of the archangel and with the trumpet call of God, and the dead in Christ will rise first. ¹⁷ After that, we who are still alive and are left will be caught up together with them in the clouds to meet the Lord in the air. And so we will be with the Lord forever. ¹⁸ Therefore encourage one another with these words.

In other words when Jesus comes back to Earth but His feet do not touch the ground, those believers who have died are resurrected and those that on the Earth are 'caught up', or raptured, to meet the Lord who is in the clouds at that moment in time.

Secondly, I would like to point out that Jesus is referred to as firstfruits of those who will be raised from the dead or made alive, in I Cor. 15:20-23:

²⁰ But Christ has indeed been raised from the dead, the firstfruits of those who have fallen asleep. ²¹ For since death came through a man, the resurrection of the dead comes also through a man. ²² For as in Adam all die, so in Christ, all will be made alive. ²³ But each in turn: Christ, the firstfruits; then, when he comes, those who belong to him.

In addition, the believers are also referred to as firstfruits, in James 1:18:

¹⁸ He chose to give us birth through the word of truth, that we might be a kind of firstfruits of all he created.

Now, in Leviticus 23: 4-6, 10-11, it says:

⁴ These are the feasts of the LORD, even holy convocations, which ye shall proclaim in their seasons. ⁵ In the fourteenth day of the first month at even is the LORD's passover. ⁶ And on the fifteenth day of the same month is the feast of unleavened bread unto the LORD: seven days ye must eat unleavened bread.

¹⁰ Speak unto the children of Israel, and say unto them, When ye come into the land which I give unto you, and shall reap the harvest thereof, then ye shall bring a sheaf of the firstfruits of your harvest unto the priest: ¹¹ And he shall wave the sheaf before the LORD, to be accepted for you: on the morrow after the sabbath the priest shall wave it.

Now, we know that the Passover symbolizes Jesus as the sacrifice, who paid for our sins and that the Passover feast was fulfilled by Jesus, at the time of Passover, 2000 years ago. The Feast of Unleavened Bread symbolizes the removal of sin, as leaven usually symbolizes sin, and the people were to eat unleavened bread only, during this time and were to also remove leaven from their houses. This was also fulfilled by Jesus as He took away our sins. Now, what does the feast of Firstfruits symbolize and did Jesus also fulfill it? We are told in John 20: 1, 15-17:

20 The first day of the week cometh Mary Magdalene early, when it was yet dark, unto the sepulchre, and seeth the stone taken away from the sepulchre.

15 Jesus saith unto her, Woman, why weepest thou? whom seekest thou? She, supposing him to be the gardener, saith unto him, Sir, if thou have borne him hence, tell me where thou hast laid him, and I will take him away.¹⁶ Jesus saith unto her, Mary. She turned herself, and saith unto him, Rabboni; which is to say, Master. ¹⁷ Jesus saith unto her, Touch me not; for I am not yet ascended to my Father: but go to my brethren, and say unto them, I ascend unto my Father, and your Father; and to my God, and your God.

So, on the first day of the week, or the day after the Sabbath Mary Magdalene came to the sepulchre (tomb) and Jesus had already been resurrected. He talks to her and tells her not to touch Him as He is not yet ascended to the Father but is about to. But later that day he did not caution them against touching Him, for it says in John 20:19-20:

¹⁹ Then the same day at evening, being the first day of the week, when the doors were shut where the disciples were assembled for fear of the Jews, came Jesus and stood in the midst, and saith unto them, Peace be unto you. ²⁰ And when he had so said, he shewed unto them his hands and his side. Then were the disciples glad, when they saw the LORD.

This suggests that Jesus had ascended and presented Himself before the father between the morning, when he spoke to Mary Magdalene, and the evening when He showed Himself to the disciples, and that the waving of the sheaf of the firsfruits of the harvest on the day after the Sabbath during the Feast of

Unleavened Bread symbolizes Jesus, the firstfruits, being presented before the Father, after His resurrection, on the first day of the week. And this means that the Feast of firstfruits was also fulfilled by Jesus.

But there is another firstfruit offering in the Feast of Weeks or Shavuot, or Pentecost, for it is written in Lev. 23: 15-17:

^{15} And ye shall count unto you from the morrow after the sabbath, from the day that ye brought the sheaf of the wave offering; seven sabbaths shall be complete: ^{16} Even unto the morrow after the seventh sabbath shall ye number fifty days, and ye shall offer a new meat offering unto the LORD. ^{17} Ye shall bring out of your habitations two wave loaves of two tenth deals; they shall be of fine flour; they shall be baken with leaven; they are the firstfruits unto the LORD.

So who are these other firstfruits that are waved before the Lord during Pentecost? Two loaves of bread baked with leaven are used. We know that leaven symbolizes sin, so these firstfruits are not perfect and that they are therefore the believers of this time that are not perfect, and do sin, and need to ask Jesus for forgiveness, and to trust Him to clean them from all unrighteousness. This is explained by John in I John 1:7-10:

^{7} But if we walk in the light, as he is in the light, we have fellowship with one another, and the blood of Jesus his Son cleanses us from all sin. ^{8} If we say we have no sin, we deceive ourselves, and the truth is not in us. ^{9} If we confess our sins, he is faithful and just to forgive us our sins and to cleanse us from all unrighteousness. ^{10} If we say we have not sinned, we make him a liar, and his word is not in us.

Thus, the two loaves of bread, or fruistfruits, seem to symbolize the second firstfruits, the yet unperfected firstfruits, who follow and trust in Jesus. The fact that these loaves are waved before the Lord, therefore, means that they will be presented to the Father at this time and would, therefore, have been raptured, or taken to Heaven, in order to be presented at the Father's throne, which suggests that the rapture may occur at the time of Pentecost.

As to when true Pentecost, or Shavuot, really is, there is some confusion, as there are many calendars and the times seem to have been changed, so it is difficult to know for certain the time appointed by God. In 2018, Pentecost according to the lunar Jewish calendar should have been observed on Monday, May 21st, actually from sunset May 20th to sunset May 21st, which is 50 days after the first day of the week during the Feast of Unleavened Bread. The Book of Jubilees seems to mention that this lunar calendar causes feast days to arrive 10 days too early, but we cannot be certain of even that.

In conclusion, the scriptures seem to indicate that the believers are to be presented at God the Father's throne on Pentecost, which suggests that the rapture or resurrection of the believers may happen then just like Jesus' own presentation to the Father came shortly after His own resurrection. As to exactly when Pentecost is, it is hard to say. This is most likely something that may only be clearly understood in retrospect.

Chapter 11

242: Planet X System affecting earth: van Allen Belt particle acceleration

The evidence that members of the system of dead stars that seems to have started invading the Solar System at least 100 years ago is overwhelming and lies in the many SDO, LASCO, and STEREO spacecraft images showing these objects are in the Sun's corona, connecting with the Sun and drawing energy from it (see Article 116: Planet X Objects: unbelievable evidence and size) [1]. The Sun is now much weakened due to their presence and will most likely go completely dark in the near future (see Article 195: Stellar Cores and the dying Sun, and Article 225: Weakening Sun: SORCE radiation measurements are not all solar radiation) [2, 3] just like the earth's nearest star, often referred to as Nemesis, and also Dark Star, seems to have done (see Article 208: Incoming Dark Star) [4].

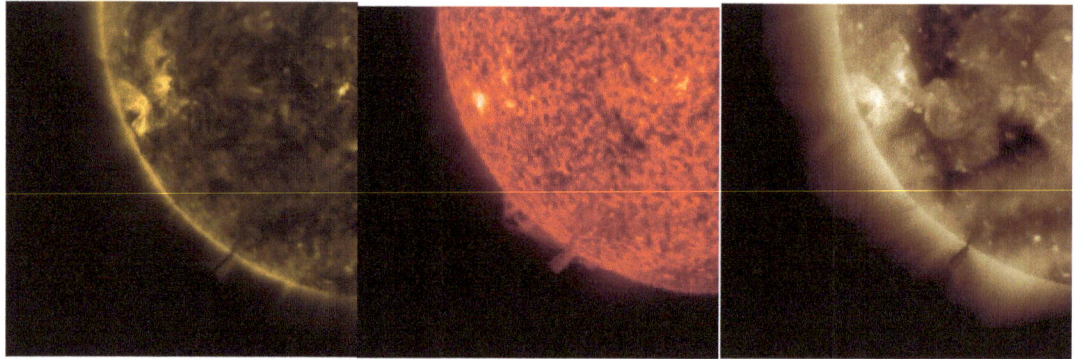

Figure 11.1. Images of the Sun, as detected by the SDO satellite, on March 11th, 2012, at 6:34 (UTC), in the 17.1, 30.4 and 19.3 nm (ultraviolet) wavelengths. A dark spherical object is seen drawing material from the Sun. The object is about half the radius of Jupiter. The dark root like the connection is not as dark in the 304 angstrom image suggesting the matter in it comes to form deep within the chromosphere [1].

When I first discovered that these objects were in the Sun's corona I thought that they would only affect the Sun but since then I have discovered that stars and planets have similar dense cores and generate energy in their cores through the same mechanism, i.e. radioactive decay or fission of unstable nuclei, and since these objects are no longer able to generate this energy in their core, they are attracted to objects that do, and will thus be attracted to planets in the Solar System as well as to the star at the center of the Solar System. And indeed the evidence that these objects are being hosted by the earth is now also overwhelming and comes from the unprecedented ocean recession events, which have occurred worldwide (see Article 188: What is causing the ocean to recede all over the world?) [5]

Figure 11.2. An empty beach, due to the ocean receding, from the Brazilian coast, on August 12th, 2017, no large storms or hurricane could be blamed for the phenomenon as there were no storms or hurricanes anyway near this coastline. This was unprecedented.

Now, NASA in an article entitled ' The case of Relativistic particles Solved with NASA mission, from May 29th, 2018 [6] has admitted that particles in the Earth's van Allen Belts are being accelerated by several means, which do not always include solar activity. The proposed reason for this is locally produced electromagnetic waves called chorus waves. However, acceleration requires an energy transformation process in the earth's magnetosphere and these are supposed to be driven by the Sun only. Waves within the earth's magnetosphere, which are able to accelerate particles, have to have a source and naming the waves does not in any way identify the source. The most likely source is, of course, the Stellar Cores which are being hosted by planet earth. These objects absorb energy in the form of photons and matter, which will contain photons inside them.

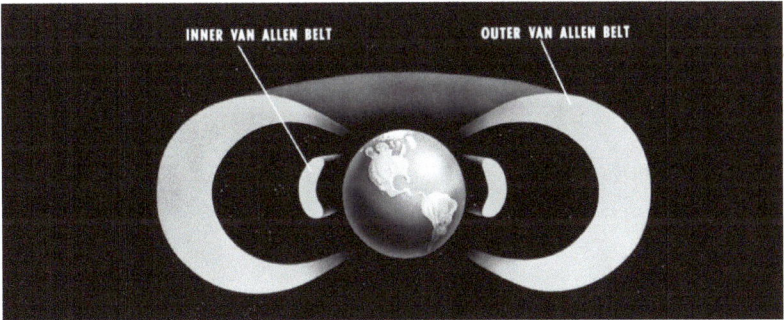

Figure 11.3. Particles in the earth's van Allen Belts are being accelerated without input from the Sun, which indicates a local source of the acceleration.

The drawing of material by these objects will cause an electrical connection to be established, between them and the earth, which will, therefore, ionize space around the objects and thus lead to the acceleration of particles in that region of space. In this way, Stellar Cores in orbit around the earth will cause increased ionization of the earth's ionosphere and accelerate particles in the Sun's van Allen Belts, without any input from the Sun through the Solar Wind.

Figure 11.4. An object which seems to be emitting red light, and is surrounded by a diffuse cloud, is seen here in a European webcam. It was caught by Jeff P in early March 2018. The object did not move across the sky but remained in the same position for an extended period of time. Stellar Cores are also often observed to stay stationary with respect to a point on the Sun with which they have made a matter connection with (see Article 227: Stellar Cores affecting the earth and possible connection to Volcanic Eruptions) [7].

In conclusion, there is growing evidence that Planet X System Stellar Cores are being hosted by and affecting the Earth in various ways, from causing unprecedented tidal events to causing unprecedented volcanic eruptions. They seem to also be the source of the electromagnetic waves causing acceleration of particles in the Earth's van Allen Belts.

References:

[1] Albers, C. (2018). Article 116: Planet X Objects: unbelievable evidence and size.
[2] Albers, C. (2018). Article 195: Stellar Cores and the dying Sun.
[3] Albers, C. (2018). Article 225: Weakening Sun: SORCE radiation measurements are not all solar radiation.
[4] Albers, C. (2018). Article 208: Incoming Dark Star.
[5] Albers, C. (2017). Article 188: What is causing the ocean to recede all over the world?
[6] https://www.nasa.gov/feature/goddard/2018/the-case-of-the-relativistic-particles-solved-with-nasa-missions
[7] Albers, C. (2018). Article 227: Stellar Cores affecting the earth and possible connection to Volcanic Eruptions

Chapter 12

243. Earth hosting at least 3 Planet X System Objects

As I have said in many previous articles the evidence that the Planet X System of Stellar Cores has not only attached themselves to the Sun but also the earth is now overwhelming, I have written extensively about their effects on the Sun and how the Sun is weakening and going dark as a result of their presence. The evidence for their presence is now mounting but the one irrefutable piece of evidence that they are in orbit around the earth is the numerous ocean recession events that have occurred around the world (see Article 188: What is causing the ocean to recede all over the world?) [1]

Figure 12.1: SDO image in the 171 angstrom wavelength from October 13th, 2017 showing a dark Stellar Core, which appears to be about half of the size of Jupiter making a matter of connection with the Sun.

Figure 12.2. Ocean recedes leaving boats sitting on mud, in the harbor in Punta del Este, Uruguay, on August 11th, 2017. The ocean came back but this extreme low tide had never happened before.

Only an object exerting a stronger tidal force than the moon could possibly cause such severe ocean recession which must be tidal in nature as the water returns many hours later just like it does with a normal tide. These objects have a weak gravitational effect for their huge mass which seems to be as a result of their inability to generate energy in their cores. This seems to be the reason why they are attracted to objects that can still generate that energy in their cores and this is the reason they are attracted to the Sun and its planets. It is likely that every single planet in the Solar System has gained some members of the Planet X System as satellites. Jupiter seems to have captured quite a few as its number of satellites seems to have gone from 16 to 69, but most likely this number will keep on going up as more and more of these objects seem to be coming into the Solar System. They start out looking black in SDO images but eventually start emitting light so the fact that new dark ones keep being observed suggests that new ones are arriving all the time.

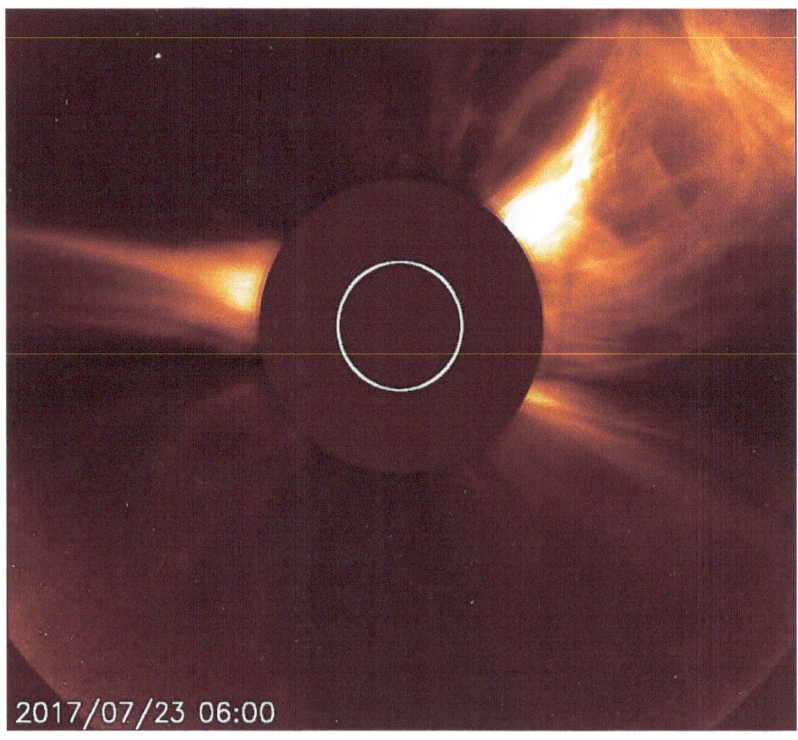

Figure 12.3. Stellar Core in a LASCO C2 image from July 23rd, 2017 moving away from the Sun within a CME. It must be within the Sun's outer corona. A size comparison with the Sun reveals that it must be about the same size as the Sun.

There seems quite a few of these Stellar Cores that have attached themselves to the Earth as the photographs shown below will attest. In the first photograph, we see one that was at that time emitting red light, they emit light because they connect electrically to the earth and thus the matter they draw from the earth forms a cloud of ionized material around them much like the coma of a comet. This seems therefore to be the reason why they look like they are enveloped in a cloud which gives off the light in a certain color. Comets also emit a lot of visible light because they discharge the solar capacitor, in other words, they emit light as a result of the Sun's generation of energy (see Article 170: Comets, planets and crustal displacements) [2]. Stellar Cores seem to emit light in a similar way to comets; they absorb matter from the Sun or the planets. This matter forms a new gaseous outer layer around the object. This outer layer, therefore, becomes a new atmosphere. The Stellar Cores also seem to shed the outer layers of material they arrive at the Sun's corona and may, therefore, do the same when they make contact with the earth's outer negative layer, such as the Earth's outer van Allen Belt. Once they have drawn enough matter and thus gained a certain amount of energy from their host objects, they start emitting light from the new atmosphere they have acquired, at the expense of the Solar System object hosting them. However, Stellar Cores would not draw current from the Solar Capacitor when they first enter the Solar System because they are so depleted in electrons that they do not have an outer negative layer like normal celestial bodies have, and it is only once they have attached themselves to a solar system body that they are able to draw current from the solar capacitor.

Figure 12.4. An object which seems to be emitting red light, and is surrounded by a diffuse cloud, is seen here in a European webcam. It was caught by Jeff P in early March 2018. The object did not move across the sky but remained in the same position for an extended period of time. Stellar Cores are also often observed to stay stationary with respect to a point on the Sun with which they have made a matter connection with (see Article 227: Stellar Cores affecting the earth and possible connection to Volcanic Eruptions) [3].

Figure 12.5. Image obtained from a video by the Youtube channel Jeff P [4]. The image comes from a web camera over Germany from October 31st, 2016. Three light sources can be seen in the image. The top one is orangey pink, the middle, and brightest, is white, edged by pink light, and the lower one is white. Chemtrail clouds in front of the objects show that they are real objects in the sky.

The fact that the objects shown above can be seen to be behind chemtrail clouds indicates that they cannot be lens flares. None can be the moon as the moon cannot be seen when so close to the Sun's position. It is possible that the brightest object is a Sun simulator and in front of the real Sun but none of the objects can be the Sun as the real Sun looks yellow from inside the Earth's atmosphere and none of the objects is yellow. We thus seem to have two Stellar Cores being observed from the Earth's surface.

Figure 12.6. Several more frames, from the same day, and the webcam as the image in figure 5. These frames show the Sun simulator move as the other two light sources remains stationary. The fact the two smaller bright sources remain stationary they are not attached to the Sun as if they were they would seem to move with the Sun in the sky.

The objects in the photographs above cannot be lens flares because they are behind a cloud, they are not the same color as the brightest light source and they do not move with the light source. A lens flare is always in front of the background objects because it is created by the lens. A lens flare is not as bright as the main object but it is the same color. As this is a web cam which is thus not moving the lens flare would have to move with the object creating it. Therefore the objects seen in figure 6 are real objects, which are, although not nearly as bright as the Sun simulator, also bright light sources in the sky.

The Sun, the moon, the planets and the stars appear to move across the sky from the earth's surface because of the earth's rotation. Any Stellar Core in orbit or even if stationary, with respect to a point on the Sun's surface would appear to move with the Sun, across the sky from Earth. Even if the largest light source is not the real Sun but a device simulating the Sun, it would be expected to be moving at the same speed as the real sun across the sky. The fact that these objects appear to remain stationary in the sky shows that they are not attached to the Sun, but must be attached to the Earth.

Figure 12.7. An object attached to the Sun would appear to move with the Sun across the sky, for an observer on the earth's surface.

Now, of the objects that orbit the earth, only artificial geosynchronous satellites are supposed to be able to remain stationary with respect to a point on the earth's surface. However, Stellar Cores are known to remain stationary with respect to the Sun's surface in the Sun's corona as they appear to become a part of the Sun when they make a matter connection with the Sun which looks like a vortex of particles being drawn from the Sun toward the Stellar Core. It, therefore, seems that the two light sources which remain stationary whilst the Sun simulator moves, are Stellar Cores, and since these are emitting a very different color of light from the object in figure 4, that we, therefore, have observational evidence of at least 3 Stellar Cores, which have made earth their host.

Since these objects are initially dark and since once attached they are not likely to disconnect from their host body, the fact that these objects are bright light sources indicates that they have been here for quite a long time and have absorbed a lot of energy from the earth. These objects will have a pronounced effect on the earth, which is likely to increase with time, and as these objects continue to gain energy. If the number, of objects, hosted by the earth, keeps increasing, this effect will increase even more. Tidal effects, severe storms, and volcanic eruptions are some of those effects (see Article 237: Hawaiian Island and Mega tsunamis) [5]

In conclusion, there is overwhelming evidence that Planet X System Stellar Cores are found in the inner solar system and both the Sun and the Earth are hosting these objects and thus being affected by them. There is observational evidence that at least 3 Stellar Cores have made Earth their host. These objects' effects on the earth range from causing earthquakes, breaking up the earth, creating tidal events, and causing volcanic eruptions and strange weather events.

References:

[1]	Albers, C. (2018). Article 188: What is causing the ocean to recede all over the world?
[2]	Albers, C. (2018). Article 170: Comets, planets and crustal displacements.
[3]	Albers, C. (2018). Article 227: Stellar Cores affecting the earth and possible connection to Volcanic Eruptions.
[4]	https://www.youtube.com/watch?v=sjcg7vCQFtA&t=438s
[5]	Albers, C. (2018). Article 237: Hawaiian Island and Mega tsunamis.

Chapter 13

244: The Planet X System: destroyer of Star Systems

As I have detailed in many previous articles the evidence that the Planet X System has invaded the Solar System and that both the Sun and the earth and most likely all the planets in the Solar System are now hosting and thus being drained of energy by these objects is overwhelming. There are by now many hundreds if not thousands of photographs of these objects within the Sun's corona (see Article 116: Planet X Objects: unbelievable evidence and size) [1]. These objects make connections with the Sun and draw energy from the Sun. It is obvious that the Sun is getting increasingly weaker as a result of these objects' presence in the Solar System (see Article 195: Stellar Cores and the dying Sun) [2]. The evidence that they have attached themselves to the earth is also overwhelming (see Article 243: Earth hosting at least 3 Planet X System Objects) [3]. In addition, these objects come in many different sizes, some are larger than the Sun and some appear to be about the same size as the Earth or even smaller.

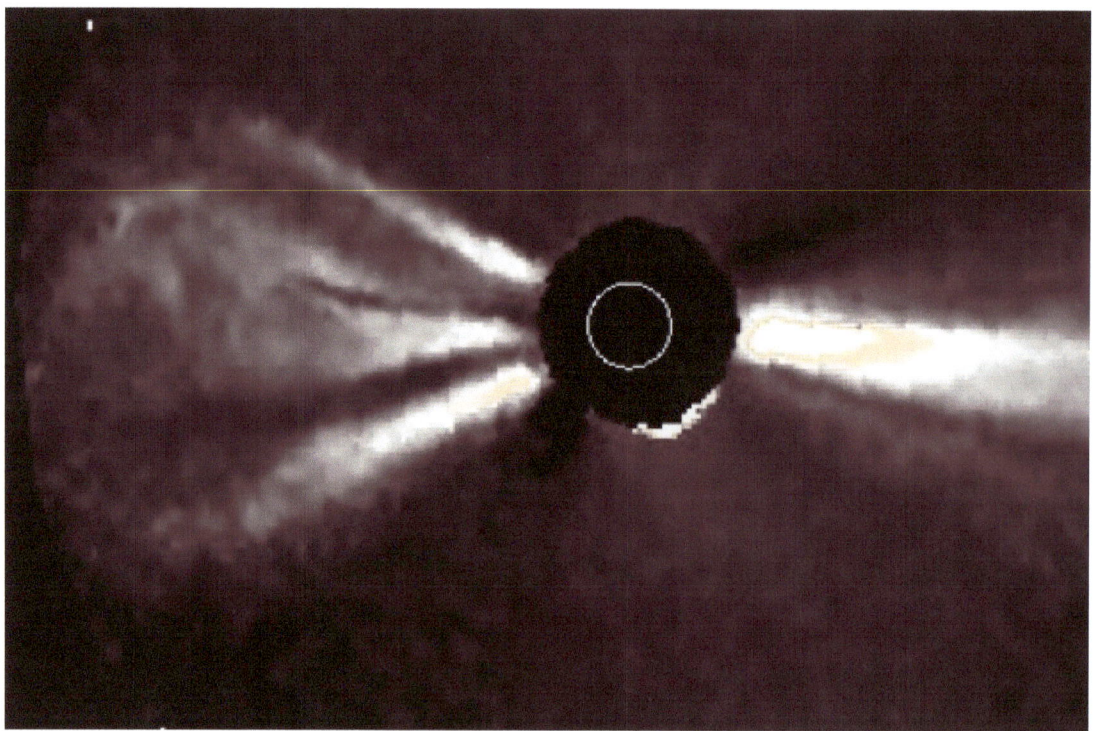

Figure 13.1. Huge Stellar Core within a CME in a Stereo COR2 image from September 13th, 2017 at 7:11 (UTC). The object appears to be at least 3 times the size of the Sun [1].

A few of the larger ones are characterized according to size in Article 116 and a diagram illustrating this appears below. Just these few objects range in size from 0.4 times the size of Jupiter to 30 times the size of Jupiter. But very small Stellar Cores are also present and in very large numbers. There is evidence that these were already present in 1996.

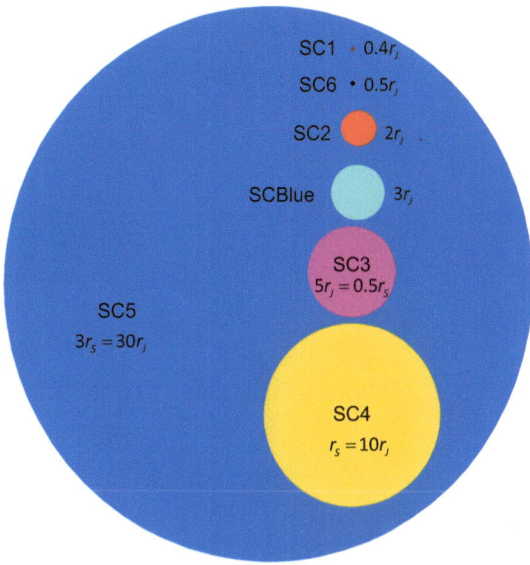

Figure 13.2. Size comparison of different Stellar Cores appearing in the Sun's corona [1].

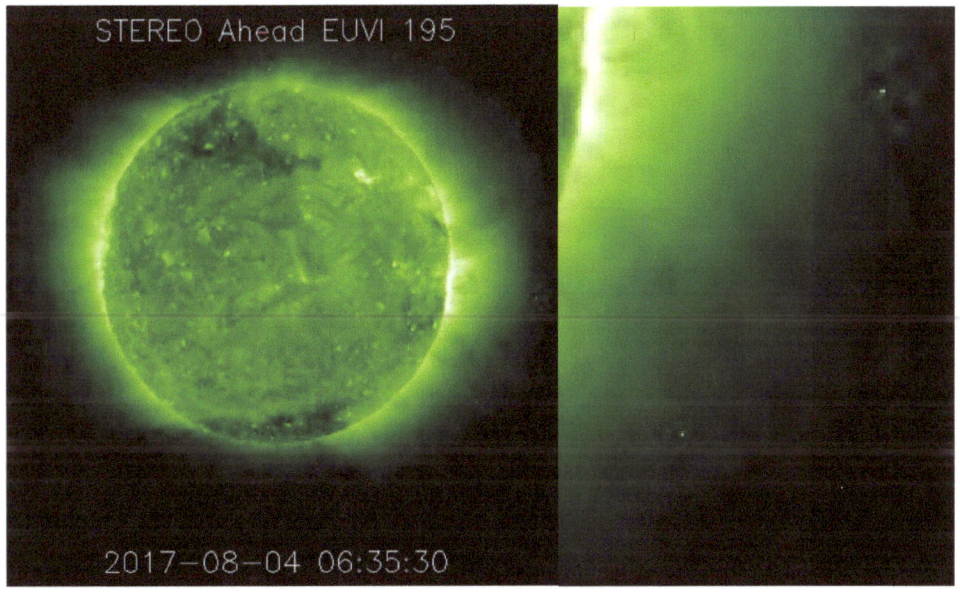

Figure 13.3. Stereo A EUVI 195 angstrom image from August 4th, 2017 at 6:35 (UTC). On the right a close up view of right hand side of a full image showing groups of clustering Stellar Cores in the Sun's corona. A white dot indicates a stellar core that has rejuvenated to the point it can now emit ultraviolet light the black dots indicate stellar cores that have not yet rejuvenated and so are newer arrivals at the Sun's corona. These objects are about the same size as the Earth.

The figures above show only some of the objects that have been observed, there must be many more presents in the Solar System. There appear to be 100s if not thousands of small Stellar Cores in the Sun's corona, in the 1996 image, shown in figure 5 below.

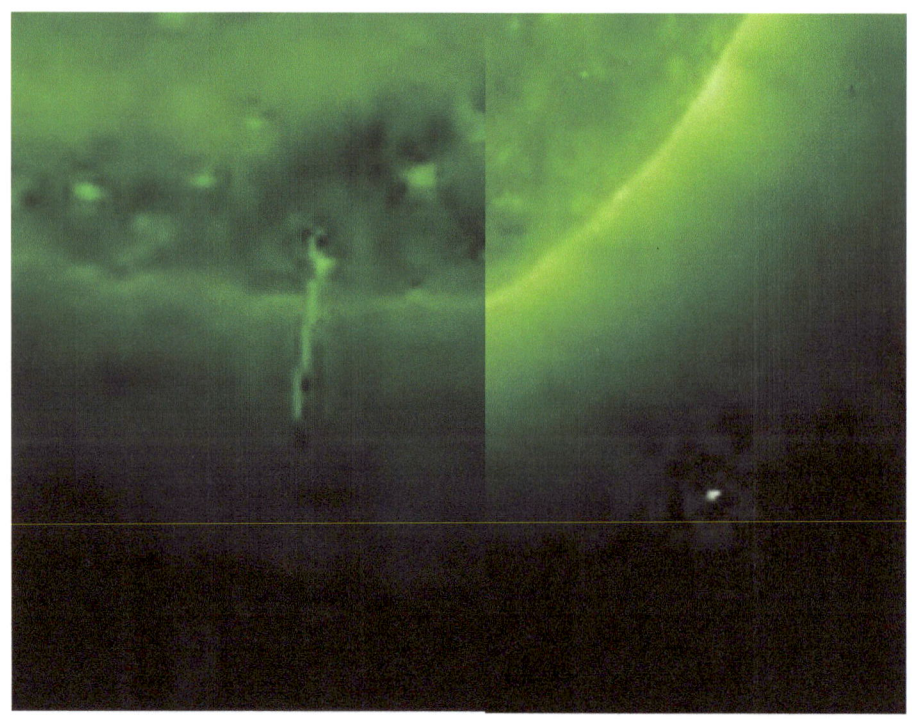

Figure 13.4. Stereo A EUVI 195 angstrom close up images of the Sun from August 4th 2017 showing Stellar Cores connecting to the Sun via plasma discharges and in the Sun's corona drawing plasma from the Sun. These objects are also about the same size as the earth.

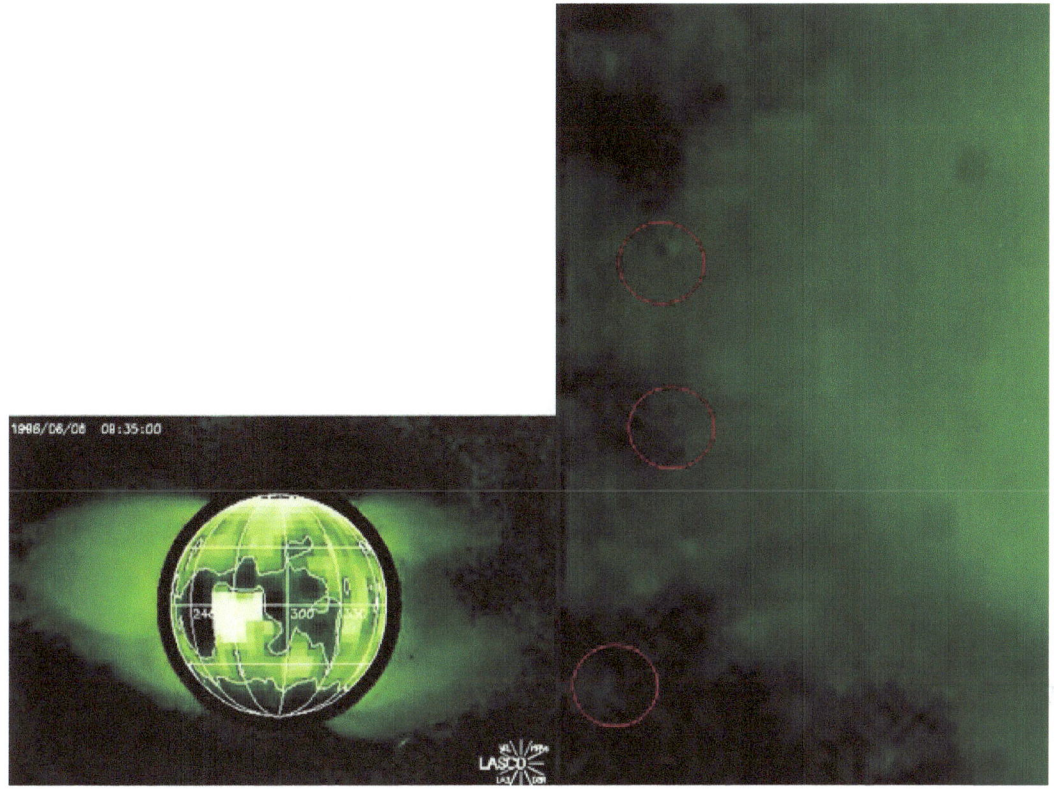

Figure 13.5. LASCO C1 image from August 6th, 1996, showing that there was already a huge number Stellar Cores, in the Sun's corona then. Only a few or highlighted above, these seemed to be very small in size (see Article 48: The lost Coronagraph and Brown Dwarf Stars Destroying a Star) [4].

Since only Stellar Cores positioned along the Sun's eastern and western limb can be seen in these images and since it is almost impossible to see those in front of the Sun, it should quickly become obvious that those that are observed must be a small sample of what must really there. In fact, this appears to be a huge system with possibly thousands of members, with most appearing to be small and thus planetary sized (see Article 54: The Planet X System is here and it is huge) [5]. However, in recent years, very large Stellar Cores have been observed in the Sun's corona as well, although not in these huge numbers. And, since at least 6 have been observed in the Sun's corona, there are likely to be 3 to 10 times that number within the Solar System. In addition, newer members of the Planet X system seem to be arriving at the Sun's corona all the time. The newer ones can be recognized by the fact that they do not emit light and also appear as dark spherical objects within the Sun's corona.

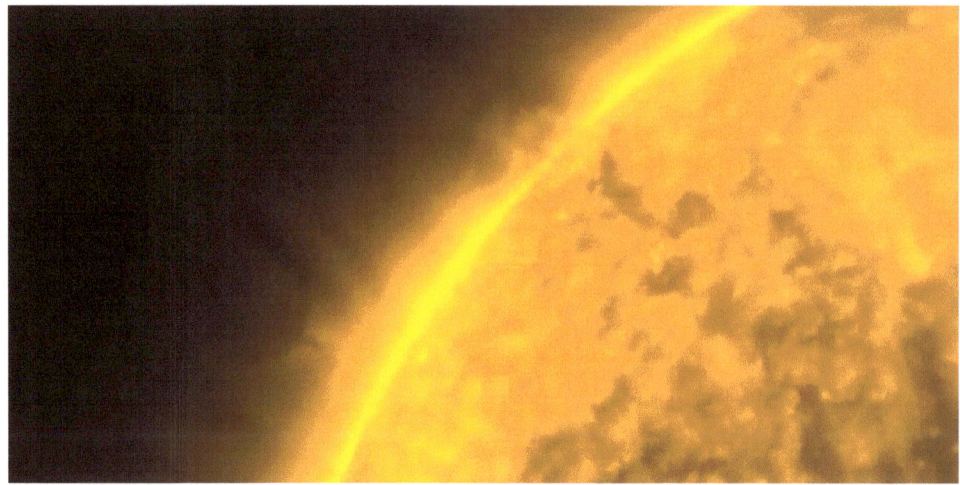

Figure 13.6: SDO image in the 171 angstrom wavelength from October 13th, 2017 showing a dark Stellar Core, which appears to be about half of the size of Jupiter.

Now, as I have detailed in many previous articles there is not much different between planets and stars. These objects seem to be formed by a galactic nucleus through the same process that causes the Sun to produce the Solar Wind and also to have CMEs (coronal mass ejections). These can be described as continuous matter creation events and as episodic matter creation events but at a much larger scale. During these events high energy photons are emitted due to electrical discharges, within a region of the high electric field and as a result, split into particles of opposite charge.

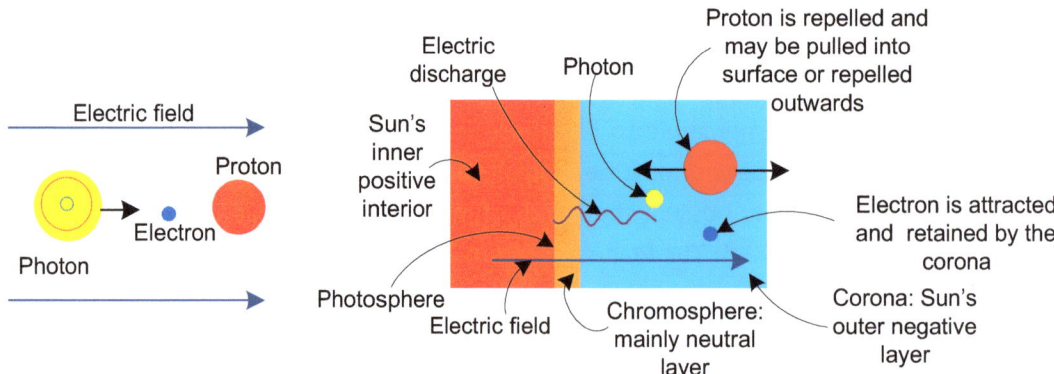

Figure 13.7. The mechanism in the Sun leading to either the production of the solar wind or a CME: electric discharges results in photon emission which then splits into particles. The gravitational interaction gives rise to the electric field and also causes protons to be ejected from the Sun's corona thus giving rise to the Solar Wind and CMEs. The same mechanism will operate in the case of a galactic nucleus but at a much larger scale.

This particle then undergo fusion and start to combine into dense cores which will then end up as celestial objects (see Article 215: Dark matter, galactic evolution, and star formation) [6]. The largest turn into light emitting dynamos, or stars like our Sun, and the smallest, which are not able to emit as much radiation, we refer to as planets. However, all the celestial objects with a core will generate energy in their cores, through radioactive decay of unstable nuclei (see Article 192: Neutron stars and fission as a star's internal energy source) [7]. This energy will allow the object to have a positively charged inner core and a negatively charged outer layer, which establishes an electric field between the core and the outer layer, which then causes electric discharges (lightning) in the object's outer layer or atmosphere. This occurs in all objects even though we may refer to some as planets and others as stars. The fact is that the earth, even though it is classified as a planet, also has electrical discharges, or lightning, in its atmosphere and therefore emits light from its atmosphere, in the same way, that a star, like the Sun, does but at an extremely low level of intensity, by comparisson. However, the much larger objects are able to have intense and continuous discharging, in their atmospheres, and are thus able to emit light to a greatly increased level.

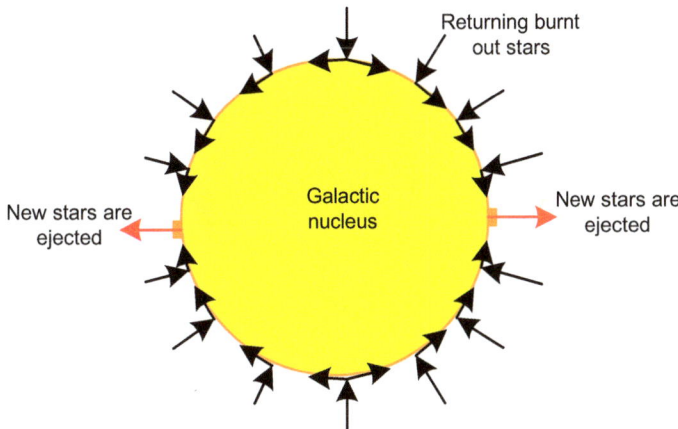

Figure 13.8. Galactic nucleus, as seen from above: New stars are ejected along two opposite directions and dead stars arrive along with all other directions.

Now, since the main difference between stars and planets is size, and since the Stellar Cores found in the Sun's corona are of widely varying sizes we realize that this invading system is not just a system of dead stars but also dead planets. In addtion, since we seem to have observed so many of these objects in the Sun's corona, with at least 6 being large enough for it to be extremely likely that they were once fully functioning stars, with most likely, planets attached to them, an image starts to emerge that the Planet X System of Stellar Cores is an agglomeration of what was once several star systems that seem to have been destroyed and consumed and absorbed by the Planet X System of dead celestial objects. The objects are dead in that they are unable to generate energy in their cores, any more, and are also so depleted in electrons that they no longer have an outer layer of electrons. It is also their lack of ability to generate energy in the core that ultimately attracts them to celestial objects that still can generate energy in their core (see Article 184: Stellar Core evolution) [8].

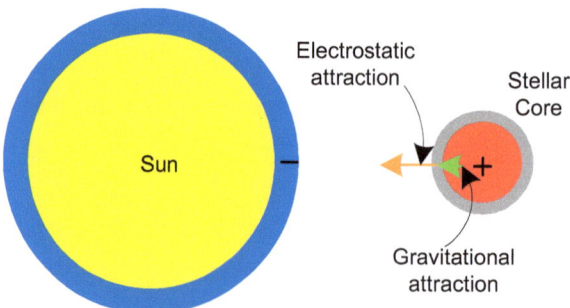

Figure 13.9. A Stellar Core as it does not have a negative outer layer acts as a super ion and is electrostatically attracted to the Sun's outer negative layer. The gravitational attraction between the two is very weak because of the object's low energy status due to its inability to generate energy in its core.

Celestial objects (stars and planets) form at the galactic nucleus' outer layer and are then ejected outwards thus forming the galaxy's spiral arms. I then thought that they would die at the outer edge of the visible part of the spiral arms, and would thus not be visible in their return journey to the galactic nucleus. But this concept of a dead star encompasses only it losing the ability to emit light. However, it has become clear that simply losing the ability to emit light is not the end, and that a star, and all planets, after losing the ability to generate energy, in their core, which may happen millions of years after losing the ability to emit any light other than infrared radiation, also lose their gravitational field and thus the ability to attract other objects, and even the ability to attract their own outer layers of matter. In addition, dying stars also lose their outer negative layer, which changes the dynamics of which objects they will then be attracted to, and how. Once the outer negative layer is lost, there will still be a positively charged inner core, and since by that time the gravitational attraction to other objects will be very low, it is the electrostatic attraction that will cause these objects to be attracted to objects that still have the ability to generate energy and thus produce a negative outer layer. Since the electrostatic interaction is much stronger (about 10^{20} times stronger) than the gravitational interaction, this would mean that these objects will be attracted from much longer distances and achieve extremely high speeds as they move toward and around objects that have attracted them.

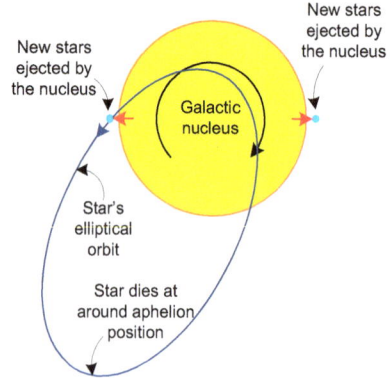

Figure 13.10. Stars ejected by the galactic nucleus have an elliptical orbit but die at around the aphelion position and are therefore no longer visible as they return to the galactic nucleus

Thus, the death process may thus cause stars to lose their way, so that instead of continuing to move in their initial orbit, which would take them back toward the galactic nucleus they will simply move on to the nearest star system that is still alive, or to the nearest star system where all objects in the system are still able to generate energy, in their core, and that are interacting through the gravitational interaction. They are however still more likely to move toward the galactic nucleus, as that would be their most likely initial direction of motion, and also, since star density will be the greatest close to the galactic nucleus and since the youngest stars and thus most energetic stars will also be found closest to the galactic nucleus, it is in the direction of the galactic nucleus that the Planet X System will be most likely to move in.

The objects in the star systems that will be attractive to the dead system may have enough energy to continue to operate in a normal way for many millions, or even billions, of years, but the dead system objects will invade it, and start to draw all the gravitational energy out of these objects that they are able to generate. As they gain gravitational energy from their host, their effects on the host will increase and they will draw more and more energy. In addition, the energy loss will cause the radioactive decay in the hosts' cores to greatly speed up so that the energy loss will keep accelerating (see Article 240: Planet X System effect on radioactive decay rate and heating of planets) [9]. As the dead system objects gain energy they will start gaining new atmospheres and start emitting light from these atmospheres but because they never regain the ability to generate their own energy, in their cores, they will simply continue to draw the energy, from the system, that is alive until no energy is left and nearly all unstable nuclei in their cores have decayed to a stable nucleus, at which time this system of objects will be dead as well, and join the Planet X System of dead celestial objects, and together, they will move on to the nearest live system. In this way, death will spread from system to system until the whole galaxy is consumed.

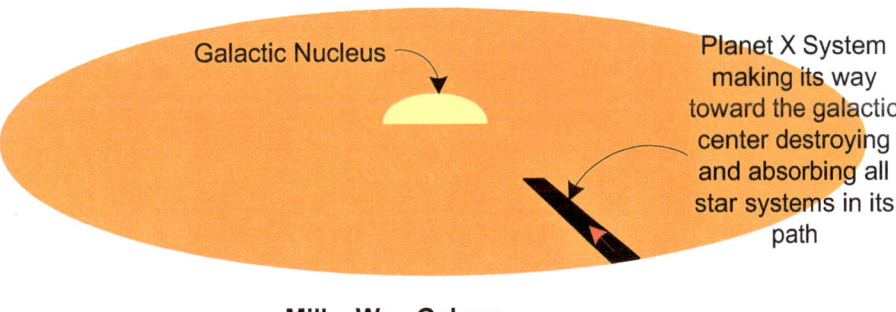

Milky Way Galaxy

Figure 13.11. The Planet X System is a destroyer of Star Systems which is likely to be making its way through the galaxy in the direction of the galactic nucleus, whilst destroying and absorbing all Star systems it encounters in its path. It thus becomes larger and more destructive with each system it absorbs.

The size and destructive effect of the Planet X System will thus continue to grow to the point that it may consume the galactic nucleus as well, and ultimately, the whole galaxy. The galactic nucleus is basically an enormous star with a radius of about 200 au, and like any other star, it will eventually become drained of energy, as well. But what is the most important to us is that it is our Solar System's turn to be destroyed and absorbed by the Planet X System. This process may take another 100 years or more to be completed. However, it is likely that it will not take nearly that long for our planet's surface to reach such a state of deterioration that it will be unable to support life.

In conclusion, the Planet X System is a destroyer of star systems. It absorbs all the energy available in a star system and then absorbs the objects belonging to that system. Thus, the Planet X System will continue to grow in size and destructive ability, as it makes its way toward the galactic nucleus. The Solar System seems to be the current system being consumed. Ultimately, this destructive process is likely to consume the whole galaxy. Our only hope is that God, the creator of the universe comes to our rescue, only He can remove these objects and thus stop them from destroying our Solar System.

References:

[1]   Albers, C. (2018). Article 116: Planet X Objects: unbelievable evidence and size.
[2]   Albers, C. (2018). Article 195: Stellar Cores and the dying Sun.
[3]   Albers, C. (2018). Article 243: Earth hosting at least 3 Stellar System Objects.
[4]   Albers, C. (2018). Article 48: The lost Coronagraph and Brown Dwarf Stars Destroying a Star.
[5]   Albers, C. (2018). Article 54: The Planet X System is here and it is huge.
[6]   Albers, C. (2018). Article 215: Dark matter, galactic evolution, and star formation.
[7]   Albers, C. (2018). Article 192: Neutron stars and fission as a star's internal energy source.
[8]   Albers, C. (2018). Article 184: Stellar Core evolution.
[9]   Albers, C. (2018). Article 240: Planet X System effect on radioactive decay rate and heating of planets.

Chapter 14

246: The ravaging of the nation of South Africa

I lived in South Africa since childhood and experienced many years of what has occurred there as a result of the change from a white government, which was judged to be repressive and racist to one which is almost completely black and yet much more repressive and unfriendly toward its people, to such an extent that it is mainly the poorest black people that have ended up suffering the most for many years. However, nowadays I believe that the suffering of white farmers has increased to such a level that these must be the most suffering people of South Africa, as these people seem to be the target of government sponsored extremely violent crime. These white farmers are also those people who feed the nation; without them, all food would have to be imported and food prices would skyrocket, thus causing, even more, suffering for the poorest of the land. But how was the pillaging of this once extremely prosperous nation started? Once they came into power, the new ANC government, which has remained in power until now, came up with minimum wage and a whole disciplinary procedure employers have to follow in order to be able to fire someone from their job, thus, people can only be fired after getting 3 warnings, within a set amount of time. This made it very difficult to fire someone, even if they were stealing and underperforming. So the result was that businesses avoided hiring people, which led to huge unemployment and crime. When they could not avoid hiring people the new standards led to low productivity and service.

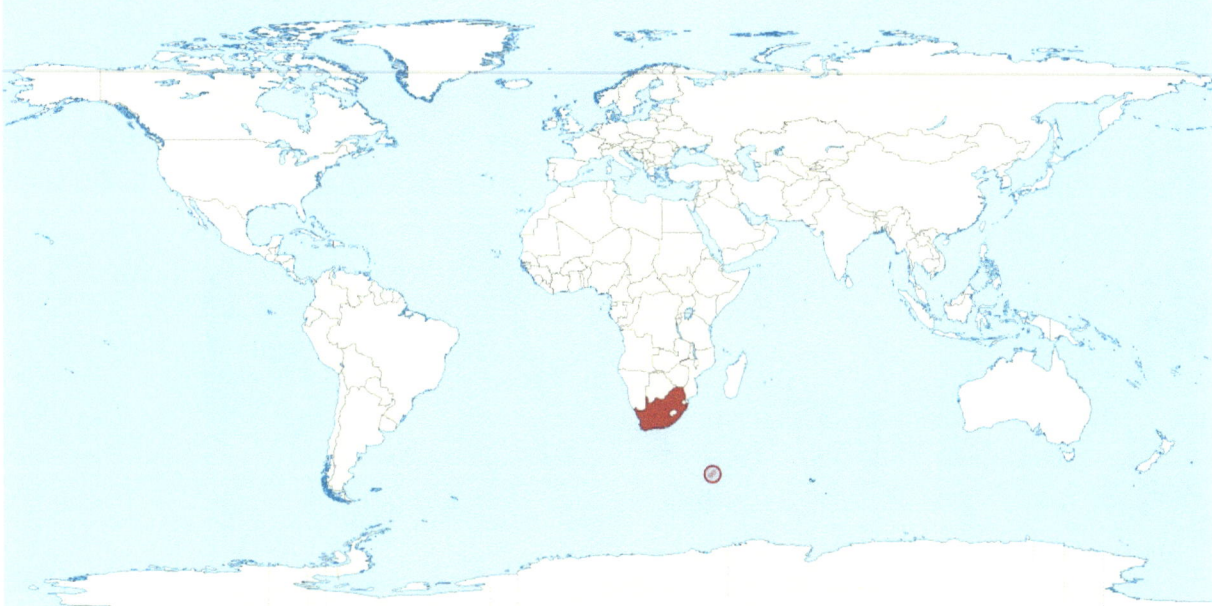

Figure 14.1. South Africa is the nation at the bottom tip of Africa.

The new government also changed the educational system to what is called outcomes based education where the responsibility for learning no longer lies with the learner but rather with the teacher. The teachers have to make sure there is some learning without causing much discomfort for the student. This means that mediocrity is encouraged and learners pass without really learning basic things, like reading and writing. Teachers make sure the students learn the minimum, which they are planning on asking them in a test later on, by giving them to fill in quiz sheets. These are sheets where the same task or question is asked several times in different ways so that by the time the student has finished filing in the quiz sheet, most of the material has somehow entered their brains. However, this has created a whole generation that does not know how to work hard or learn the material from written notes. They are often unable to memorize anything on purpose and have no idea how to go about doing that. This leads to a disastrous first year at university for many students. But the worst part of outcomes based education is that the students are encouraged to think that all questions and problems have many different and often opposing but still valid solutions, even in maths and physics, and that any of these different solutions will be equally useful. In other words, they are purposely taught not to expect there to be one standard and one correct way of doing things, which means that they cannot clearly think through a problem and get to the most valid conclusion, and are thus unable to do any real problem solving. This has created a work force, which, in general, has no work ethic, and no ability to solve problems, in their day to day interactions in their work environment, and you can see it and experience it often when you visit stores and companies in South Africa.

Figure 14.2. South Africa is beautiful. There are breathtaking beaches, mountains and wildlife reserves.

Those children that have been fortunate enough to be taught by their parents to work hard and think clearly and these are mostly children with white, Indian and colored parents, although there are also black parents nowadays that also do this great service for their children, have a great advantage, therefore. But the government got around that by demanding that employers hire job applicants from mostly the black population, which is the majority. It goes like this: when you review all the job applicants, if there is one black job applicant who is judged to be able to do the job, then he must be hired ahead of any job applicants from the other races, even if they are all better qualified than the black applicant. This has the effect of further encouraging mediocrity and low productivity, in all sectors of the society, and in addition, leaving black people in high positions feeling inferior and disempowered, because, after all, it is not really their qualifications or abilities that have been likely to get them to that position, but simply the color of their skin.

Then there is the fact that the police force has been made totally ineffectual by bypassing people, with a work ethic and motivation, for promotion, and also making their lives difficult so that they chose to leave. Thus, most of these people move to the private security industry which is thriving as a result of the very high crime rate. In addition, police training has been severely reduced so that, for example, traffic police officers do not know how to direct traffic anymore, and private insurance companies train and employ their own teams to perform that service whenever there is a power failure, and the traffic signs are not working. The power failures are extremely frequent because the power infrastructure is owned by the government, who then placed people from their political party in managing positions, and these managers decided that getting huge bonuses for many years, was much more important than maintaining and growing the power infrastructure. The lack of power has led to a curtailment of economic growth and thus the generation of jobs, as well.

Then there is the fact that hospitals, which during the previous white government were well equipped and gave all racial groups world class health care for a very tiny amount of money have been just about completely destroyed. These hospitals now have no equipment and are filthy and often have elevators that do not work ever again. The equipment was stolen by the staff, no one tells anyone else what to do so no one cleans, and there is no money to fix elevators because the money has been stolen at some level of government, so the people have no public hospitals and whatever care there is from nursing staff is usually without any respect or caring attitude toward the patients. The nursing care is so bad that even private hospitals are afflicted by the disdainful and uncaring attitude of many nursing staff. Doctors are often caring but they cannot be around all the time. And it is the poorest people in the society which are most affected by what this new government has done as most no longer can find unskilled jobs, they have no hospitals or decent education for their children.

In conclusion, the nation of South Africa has been ransacked by the new government and is another case of a nation that has been conquered through governmental change and illustrates what one expects to see when the white horse rides after seal 1 of Revelation 6 is opened.

Chapter 15

247: The effect of the Planet X System on the earth

The Planet X System is a system of dead stars and planets, which moves from star system to star system absorbing all the energy available in these systems, and thus destroying them. Once all the energy has been drained, by the Planet X System, from a living star system, all the members of this system will be dead, or energy depleted. This means that all the stars and planets of the once living system would have become Stellar Cores. The Planet X System is thus ever growing in size (see Article 244: Planet X System: destroyer of Star Systems) [1].

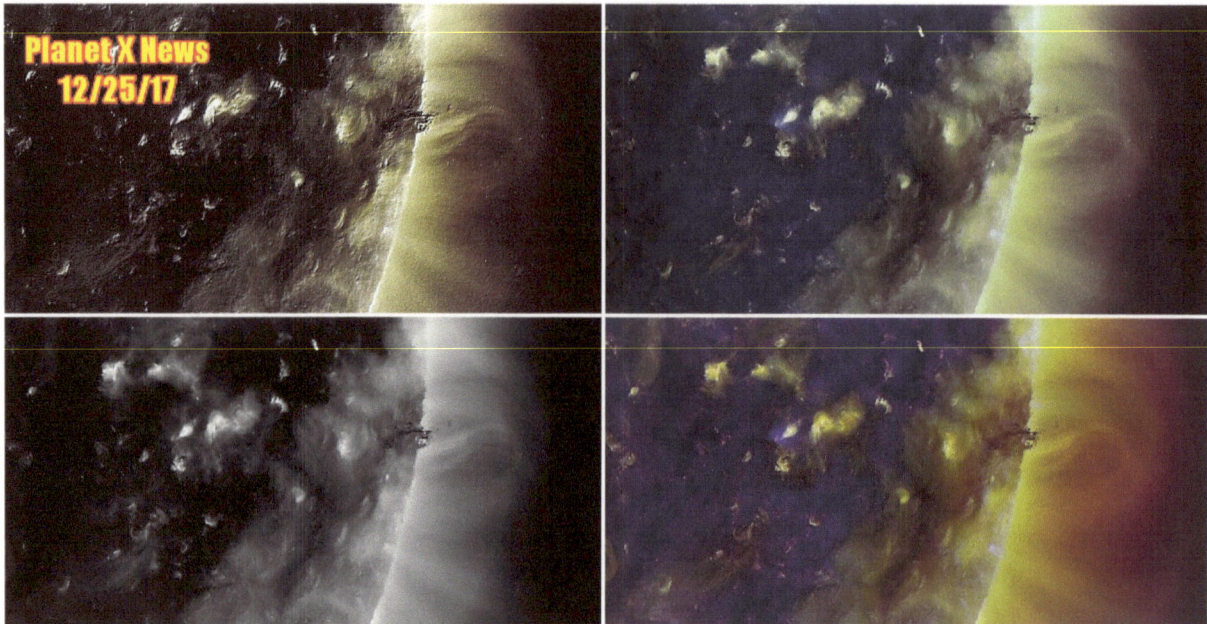

Figure 15.1. A Planet X Object or Stellar Core appears in the Sun's corona. The object is striped and a size comparison with the Sun reveals that it is about 4 times larger than the earth. The objects in the Sun's corona are usually seen in sun observing satellite images.

Figure 15.2. Another Stellar Core observed moving Away from the Sun in a Stereo A COR2 image. This object is about half the size of the Sun.

Figure 15.3: Two Planet X Objects or Stellar Cores in the Sun's corona seen in a 171 angstrom wavelength SDO image from October 13th, 2017. The dark Stellar Core is making its vortex connection with the Sun indicating that the interaction is a weak gravitational force and thus tidal in nature as it affects a small region of the Sun just below the object but not other regions further away. The objects start t emit light after being at the Sun's corona for some time so the dark object must be a recent arrival. The dark Stellar Core is about half of the size of Jupiter.

Living stars and planets generate energy through the mechanism of radioactive decay in their core. At formation, all celestial objects will have a certain amount of unstable nuclei which by decaying into lighter nuclei release energy. This initial amount of unstable nuclei in a celestial objects' core is like a reservoir of energy. It may also be visualized as a battery which is fully charged at the beginning of the celestial object's life but which runs down eventually. It can take billions of years for all the nuclei to decay and thus release the energy contained in them but eventually, the core will run low on nuclei that can still decay. Since the energy in the core is directly related to the strength of the gravitational interaction, an energy depleted object is not able to strongly attract its own particles and other objects. This causes the object to expand in size.

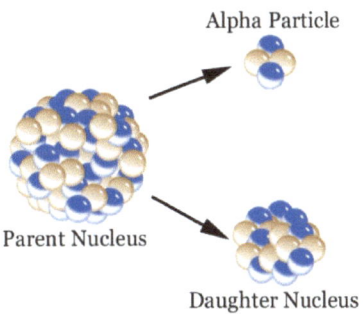

Figure 15.4. Radioactive decay is the breaking up of a heavy nucleus into 2 lighter nuclei. This releases energy or photons. Heat is a form of energy and all energy is in the form of photons which exist within

particles and cause the particles to move or vibrate in response to the amount of photon energy inside them (see Article 191: The photon universe: it is all made out of light) [2].

The gravitational interaction has two main parts, which deeply affect what then happens to the object. One of the parts is the repulsion between protons and electrons, which causes celestial objects to have a positive inner core and negative outer layer, when this interaction, which can be referred to as the charge separation part of the gravitational interaction, drops in strength, electrons move inwards toward the core and thus the electric field or electric potential generated by the object drops, which cause it to lose the ability to emit light. It also loses its outer negative layer, which becomes neutrally charged instead of negatively charged (see Article 184: Stellar Core evolution) [3].

Figure 15.4. One of the Stellar Cores attached to earth and drawing energy from the Earth (see Article 243: earth hosting at least 3 Planet X System Objects) [4]. It was caught by Jeff P in early March 2018 in a German webcam. The emission of red light occurs as a result of the objects drawing energy from the earth. The object did not move across the sky but remained in the same position for an extended period of time. Stellar Cores are also often observed to stay stationary with respect to a point on the Sun with which they have made a matter connection with (see Article 227: Stellar Cores affecting the earth and possible connection to Volcanic Eruptions) [5]. These objects can remain stationary because they are able to connect to and thus become a part of the object they are connected to and can thus rotate with the host body.

Figure 15.5. Stellar Core connected to the earth in false color. The object seems to have several layers of gaseous material around it which is emitting light and must thus be gaseous plasma. The bright straight line must be a part of the connection it is making with the earth's atmosphere, the fact that the line can be seen indicates it is emitting light and thus the gaseous making up the line is highly ionized.

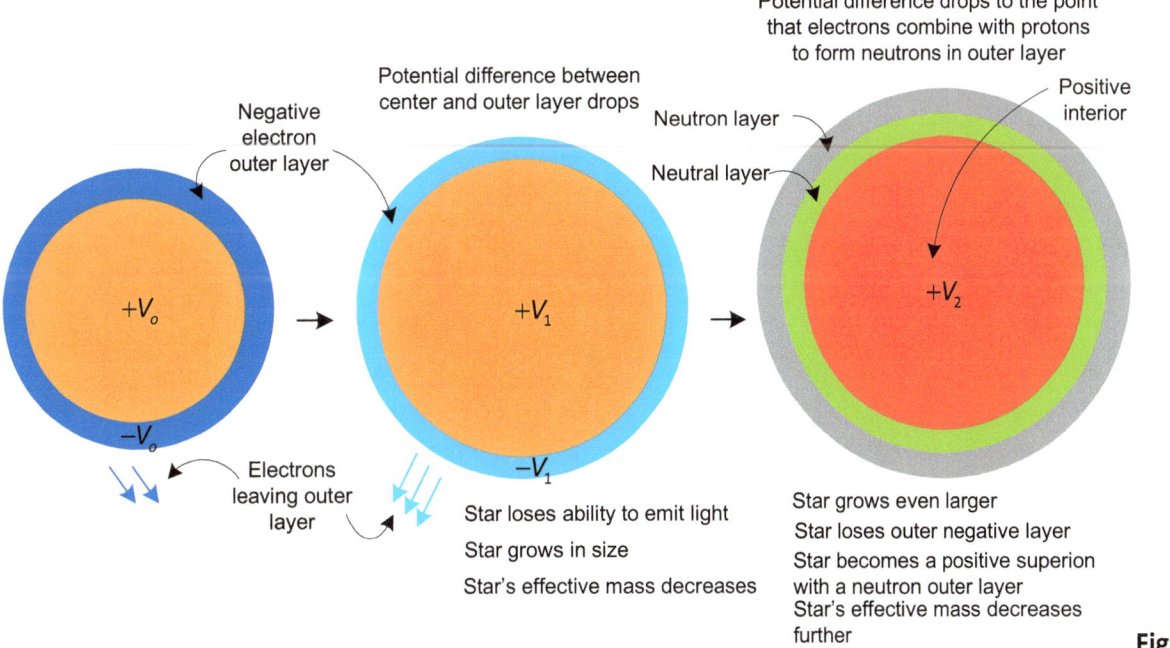

Figure 15.6. Illustration of how a star's ageing process, which causes a decrease in gravitational energy, works: Once a star's reservoir of unstable nuclei declines it loses the ability to emit light and expands in size because the strength of the gravitational interaction it is able to have decreased. At the Stellar Core stage (neutron outer layer) any electron captured by the star would most likely increase the size of the neutral layer rather than form a negative outer layer so the star would remain deficient in electrons

until it enters the Sun's corona and starts drawing them from the Sun [3]. If the drain on the Sun is extremely drastic the Sun loses its driving potential and goes dark (see Article 232: The Sun can go dark: the implications) [6].

The other important part of the gravitational attraction is the attraction between protons which gives rise to the strong force, and the force which we know as gravity. When, as a result of energy depletion, the strength of the interaction drops, the object's ability to attract, its own particles, in its outer layers, decreases, which causes the object to expand in size. The object will also lose the ability to attract any of its satellites, so a star would lose its planets and a planet would lose its moons. This object, which has now become a Stellar Core, will still retain a less positive but still positively charged inner core and will thus be attracted via the electrostatic interaction to objects which still have an outer negative layer and are thus still generating energy in their cores.

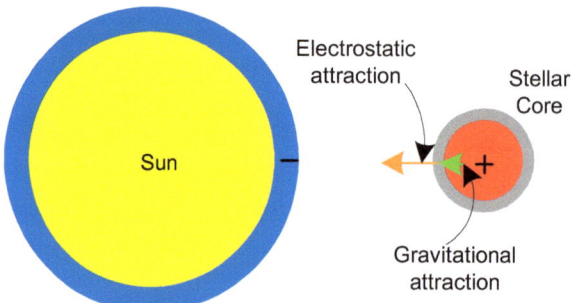

Figure 15.7. A Stellar Core is electrostatically attracted to the Sun. The gravitational attraction between the two is very weak.

Stellar cores, because their matter is so depleted in energy, absorb any energy they come across, the absorb radiation or photons, and they absorb matter which contains energy in the form of photons. It is these photons which are responsible for whatever strength the gravitational interaction is able to have. So these objects absorb light and matter. In the meantime the object, which becomes the host of the Stellar Core and is losing energy starts to lose more energy than it is generating, which causes the rate of radioactive decay in its core to increase (see Article 240: Planet X System effect on radioactive decay rate and heating of planets) [7]. This means that it starts running through its energy reservoir faster. In the meantime, the Stellar Core gains enough energy to start emitting light but all at the expense of its host. As more and more Stellar Cores arrive and start also depleting the host, it loses energy more and more rapidly, and thus, the radioactive decay rate keeps increasing in response, until the host runs out of energy and becomes a Stellar Core itself. It then moves with the other Stellar Cores in the direction of another host that is still alive, and thus still generating energy in its core.

But, before the host reaches the point when it too turns into a Stellar Core, its surface temperature would increase, because more energy is arriving at, or moving through, its surface, as a result of the increased rate of energy generation, in the core. The host would also start expanding as the strength of the attraction on its own outer layers decreases. This causes the earth to expand and its outer layers to crack open so that fissures and sinkholes appear. But the gravitational decline is likely to be unstable so that different regions are likely to experience a greater drop in attraction to the core than others and this will lead to sea levels to vary wildly over different regions thus explaining the ocean recession events the earth has been experiencing. In addition, since the earth has a hot liquid layer (magma), the breaking up of the surface causes the magma to more easily reach the surface which will lead to intense and possibly never ending volcanic eruptions, which will only increase in severity over time. Eventually, the earth's crust will be completely broken up and vast regions will be covered in continuously flowing lava.

Figure 15.8. As the earth loses energy to the Stellar Cores which causes its surface to split open or crack.

Figure 15.9. As the earth loses energy and expands volcanic eruptions will become very severe and large portions of the surface will be covered in continuously flowing lava.

As the Stellar Cores absorb energy, the strength of the gravitational attraction they are able to exert on a planet like earth will increase to the point that crustal displacements and thus pole shifts are likely to occur and these occur repeatedly, because the Stellar Cores do not leave until all of the planet's energy is gone, by which time all its rocky surface will be completely destroyed. At the end of the process, the celestial objects lose most of their outer layers of material and only the solid core and a thin layer around the core will be left.

So what will occur to Earth? Well, Earth seems to have attracted more at least 3 Stellar Cores [4] and it will most likely attract more, as time goes on. Earth has experienced sink holes for many years now and has also experienced unprecedented ocean recession events. All these indicate that earth's gravity is decreasing and doing so in an unstable manner. The fact that the Hawaiian volcanic eruption shows no sign of stopping but seems to be getting continuously worse is an indication that earth's deterioration is now reaching an advanced stage, which can only worsen as time progresses. The next step would most likely be more volcanoes starting to erupt and maybe a crustal displacement, with worldwide mega earthquakes occurring as a result. Unending volcanic eruptions will eventually lead to parts of volcanic islands undergoing volcanic activity to slip or collapse into the ocean, which then generates megatsunamis (see Article 237: Hawaiian Island and Megatsunamis) [8].

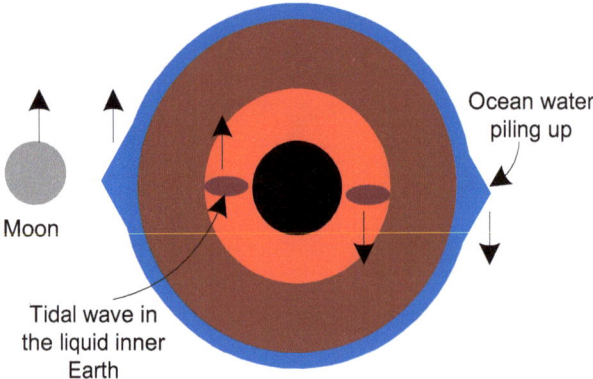

Figure 15.10. A tidal wave is generated inside the earth's liquid outer core as a result of the tidal force exerted by an object close to the Earth. This wave causes the outer solid layers to disconnect from the core and move independently which the causes the poles to end up in a vastly different positions.

Since the Stellar Cores attached to the earth cause more energy to be generated in the earth's interior, all of earth's layers will be more energized, including the ionosphere, which will lead to increased severity of storm systems and increased lightning. These are all events that we have been experiencing more and more, for some years now.

Figure 15.11. Increased severity of lightning storms in one of the consequences of having Stellar Cores drawing energy from the earth as this will greatly ionize the earth's ionosphere.

In conclusion, the Planet X System will have a very serious and disastrous effect on the Earth. Its effect is likely to keep on increasing as time progresses. Some of these effects are the appearance, or incidence, of fissures, sink holes, volcanic eruptions, crustal displacements, earthquakes and increased lightning. The earth's surface and oceans will also increase in temperature as unstable nuclei release more energy due to an increased rate of radioactive decay, which causes more heat to be produced in the core. This heat flows outwards from the core and moves through all of the earth's layers. This, in turn, will heat the surface of the earth, atmosphere, and oceans and it will also heat the magma, in the earth's interior, and thus also greatly contribute to the increase in volcanic eruptions.

References:

[1]   Albers, C. (2018). Article 244: Planet X System: destroyer of Star Systems.

[2]   Albers, C. (2018). Article 191: The photon universe: it is all made out of light.

[3]   Albers, C. (2018). Article 184: Stellar Core evolution.
[4]   Albers, C. (2018). Article 243: Earth hosting at least 3 Planet X System Objects.
[5]   Albers, C. (2018). Article 227: Stellar Cores affecting the earth and possible connection to Volcanic Eruptions.
[6]   Albers, C. (2018). Article 232: the Sun can go dark: the implications.
[7]   Albers, C. (2018). Article 240: Planet X System effect on radioactive decay rate and heating of planets.
[8]   Albers, C. (2018). Article 237: Hawaiian Island and Megatsunamis.

Chapter 16

248: Planet X System Objects Interacting with earth

A screenshot from a video by the Youtube channel 'Pieasand cakes' is shown below [1]. Two bright objects leaving a trail behind them are seen moving across the sky. The Sun is to the right and just off the frame. The objects are moving in the opposite direction to which the Sun would move in the sky. Since it is extremely difficult to observe objects, which are outside of the earth's atmosphere during daylight hours, from the earth's surface, especially with the current haziness of the earth's atmosphere due to chemtrails, and since the objects are, in addition, quite close to the Sun's position, it is likely that they are moving inside the earth's atmosphere.

Figure 16.1. Two objects streak across the sky in a direction which is opposite to the earth's rotation and are thus in space and moving much faster than it is possible for any object to move or they are inside the earth's atmosphere.

Now, the objects are moving with the same speed, and in parallel trajectories, and are thus moving together, which indicates that they are a binary. But in binaries, which interact via a normal strength gravitational attraction, we usually have the less massive object orbiting the more massive object unless the binary formed recently from one object splitting into 2. However, in that case, they should be following the same trajectory with one simply being slightly in front of the other. These objects are following parallel side by side trajectories instead, which suggests that they are not ordinary objects and are not attracted to each other, or the earth, through the normal gravitational interaction, which immediately suggests that they are part of the Planet X System of Stellar Cores (see Article 244: The Planet X System: destroyer of Star Systems) [2].

.

The objects are very bright and thus emitting light and cannot, therefore, be airplanes and therefore these are not chemtrail planes. They cannot be meteors as meteors streak across the sky and impact the ground very quickly whereas these two objects are flying at a constant altitude with a trail behind them. They are therefore most likely very small Stellar Cores inside the earth's atmosphere. Stellar Cores interact through a very weak gravitational field and are often observed hovering over a point on the Sun's surface, and they tend to get very close to the Sun's surface (within the corona or the Sun's atmosphere) but yet do not impact the Sun. Stellar Cores may also be able to follow a parallel trajectory with respect to their host body as the 2007 Stellar Core traversed the Sun in the Sun's corona, and thus maintained an approximately constant distance from the Sun's surface as it moved inside the corona or the Sun's atmosphere.

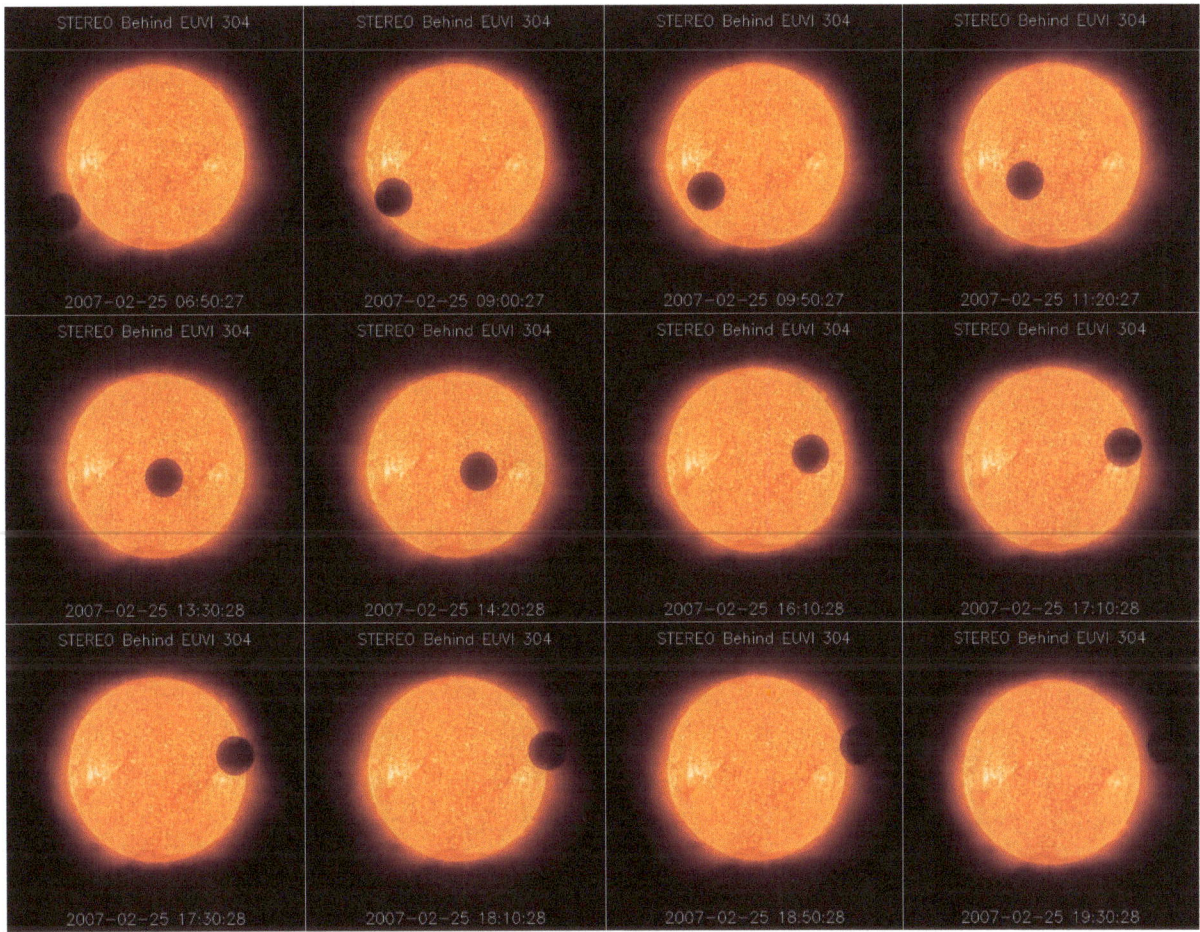

Figure 16.2. Stereo B EUVI 304 angstrom wavelength image from 2007: the 2007 Stellar Core traverses the Sun in the Sun's corona. The object was 2.2 times the size of Jupiter (see Article 143: Planet X traverses the Sun: irrefutable evidence) [3]. An object further from the Sun than Sun's inner corona would loop black in the images. This object looks and as it is entering and leaving the corona, the part of the object outside the corona looks black.

In addition, the fact that the two objects in figure 1 can be seen so clearly suggests that they are close to the observer on the earth's surface and therefore inside the earth's atmosphere. Since they do not behave like objects which strongly interact through the gravitational force, they have to be Stellar Cores and the fact that they are bright and have a trail indicates that they are drawing material from the earth's atmosphere. These objects are most likely bright for the same reason that a comet develops a coma: ionization of gases; ions recapturing electrons is what causes the plasma to emit light (see Article 169: Planetary formation: comets to planets) [4]. The objects are obviously emitting light from their comas (atmosphere around the nucleus of a comet). This shows that the objects are electrically charged or are charging up in the earth's atmosphere. Since the earth's ionosphere is the most charged part of the earth's lower atmospheric layers, it is possible that the objects are moving through the earth's ionosphere. Since the objects seem to be very small in comparison to the size most Stellar Cores seem to have, it is possible that either these are just extremely small planetary or moon Stellar Cores, i.e. the core part or what is left of what was once a live planet or moon, or they are Stellar Core debris.

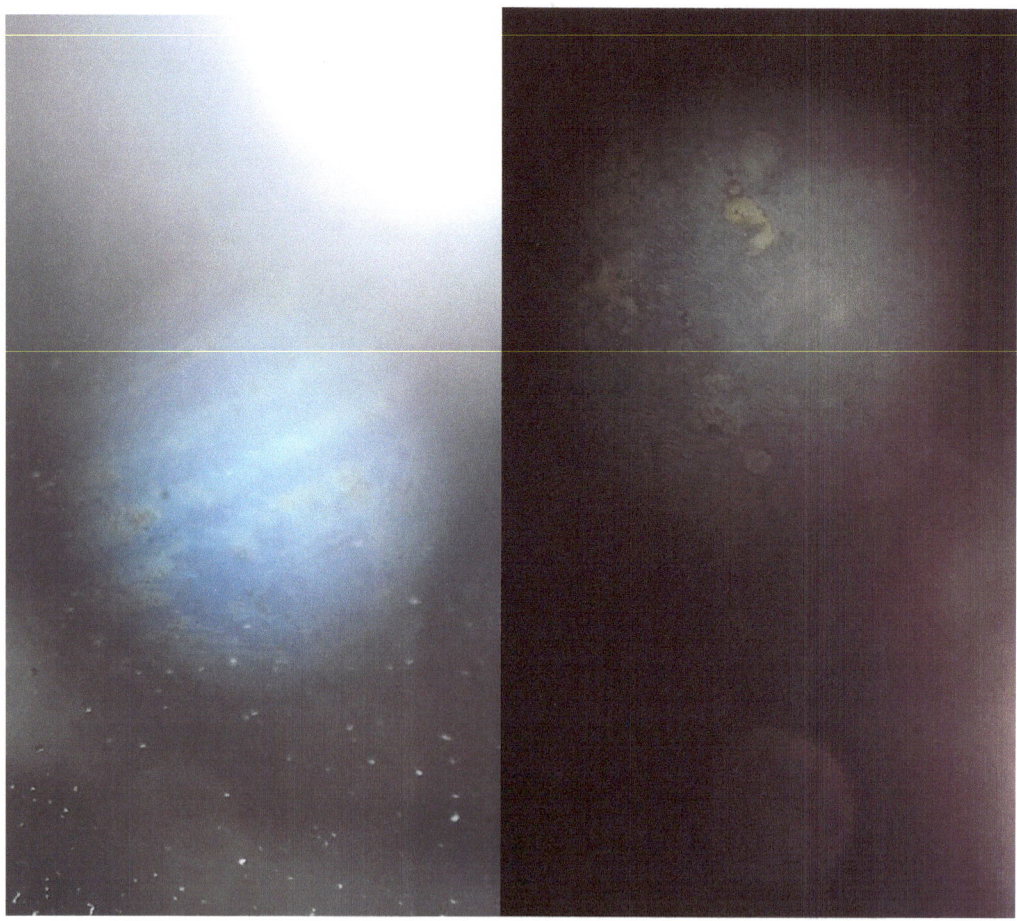

Figure 16.3. On the left: the Telescopic image of the Blue Stellar Core from May 12[th,] 2017 showing that the material making up its stripes is solid but clumpy as it is obviously being shed in small clumps. The debris resulting from the shedding of this material floats around the object. The debris is obviously not affected by any gravitational interaction indicating that it has zero photon or gravitational inside it. On the right: Photographic image of the same object from July 26[th,] 2017 showing that it had shed almost all the remaining material clinging to its surface.

The Stellar Cores are known to shed their outer layers as the Blue Stellar Core shed outer layers of material which gave it a striped appearance, within a few months. In addition, it has become obvious that the inner Solar System is full of debris floating around and looking like either black or white dots in satellite images. These spots are observed moving away from the Sun and seemingly being buffeted by a wind whenever the Sun has a CME which suggests that they have zero gravitational energy.

Figure 16.4. Stellar Core in a Stereo COR2 image from November 30th, 2017. The object is approximately half the size of the Sun. Debris (black and white specks) can also be seen.

The strength of the gravitational attraction an object is able to exert on other objects is dependent on how much photon or gravitational energy is inside the particles that make it up (see Article 210: Stellar Core gravity: tidal and G is not constant) [5]. The shed debris was seen above obviously has zero gravitational energy, so in order for the objects moving through the earth's atmosphere seen in figure 1, to be Stellar Core debris, they would have had to absorb gravitational or photon energy from the earth so that eventually, they would be able to weakly attracted other celestial objects. In this case, they would start orbiting a celestial object from afar at first and then slowly spiral in toward the object, i.e. the orbital radius would slowly decrease.

Figure 16.5. Very small Stellar Cores or Stellar Core debris that has gained some gravitational energy moving inside the earth's atmosphere, drawing energy and matter from the earth.

In conclusion, two objects moving through the earth's atmosphere in a path parallel to the earth's surface, and with colored gaseous trails behind them, appear to be either very small planetary Stellar Cores or Stellar Core debris that has become attached to the earth and are drawing energy from the earth and matter from the earth's atmosphere.

References:

[1] https://www.youtube.com/watch?v=MlVyL3ZNuNw
[2] Albers, C. (2018). Article 244: The Planet X System: destroyer of Star Systems
[3] Albers, C. (2018). Article 143: Planet X traverses the Sun: irrefutable evidence.
[4] Albers, C. (2018). Article 169: Planetary formation: comets to planets.
[5] Albers, C. (2018). Article 210: Stellar Core gravity: tidal and G is not constant.

Chapter 17

249: The photon field: Gravity, antigravity and gravitational energy

In this article, I describe with the help of diagrams what gravity and antigravity are, and what gravitational energy is. Particles come from photons. The photon splits into two separate particles of opposite charge: a proton and an electron. The proton has a much higher mass than the electron. The mass is a form of energy. Photon energy is converted to mass energy when the two particles emerge. The rest of the photon energy ends up inside the particles and is equally shared between the two. The gravitational field generated by each object is dependent on the mass of the particle and the amount of photon energy inside each particle. Gravity, therefore, owes its existence to the photon and could, therefore, be called the photon field. Gravity also occurs in two forms, the one we call gravity and we know the most about is associated to protons, so I refer to it as the proton gravitational field, which is usually much larger than the other type of field, the electron gravitational field. Electron gravitational fields are repelled by proton gravitational fields as they have opposite polarity and can thus be thought of as antigravity fields.

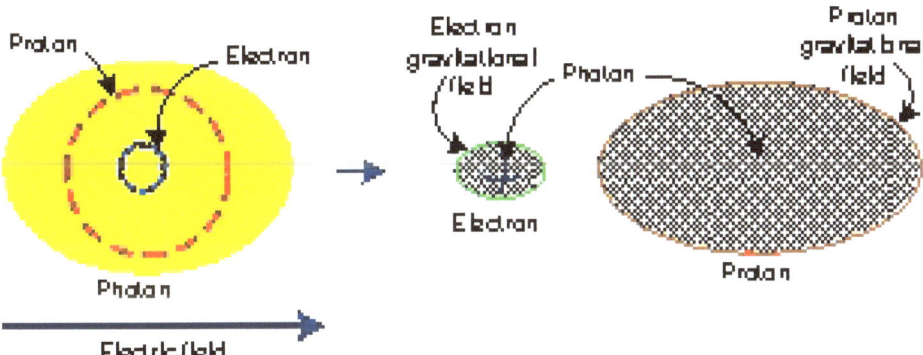

Figure 17.1. A photon splits into a proton and an electron when moving through a region of high enough electric field. The photon's energy goes into giving the two particles mass, the remaining energy is shared equally between the two particles and exits as particle confined photon energy, or gravitational energy, as this energy is directly connected to the strength of the gravitational interaction the particle will be able to have. The proton and the electron have gravitational fields with opposite polarity so that there is a repelling gravitational force between them. The size of a particle's gravitational field corresponds to the magnitude of the force it is able to exert on another particle.

The mass of a particle is a measure of the amount of photon or gravitational energy it is able to hold. The strength of the gravitational interaction a particle, or object, is able to have is given by how far beyond the edge of the object, the gravitational field extends. If it extends far beyond the edge of the object than the object is able to interact strongly, but if it does not extend very far, than the object will interact weakly.

The strength of the interaction is associated with the amount of photon energy in each particle. Upon making contact particles share their photon energy equally. This is a very important concept in this theory, as it allows us to understand how Stellar Cores draw energy from living (high photon energy) celestial objects. Celestial objects act in a similar way to particles so that Stellar Cores when coming very close to a normal celestial body, will draw the object's photon, or gravitational energy, and thus increase the size of their gravitational fields whilst decreasing the size of the host body's gravitational field.

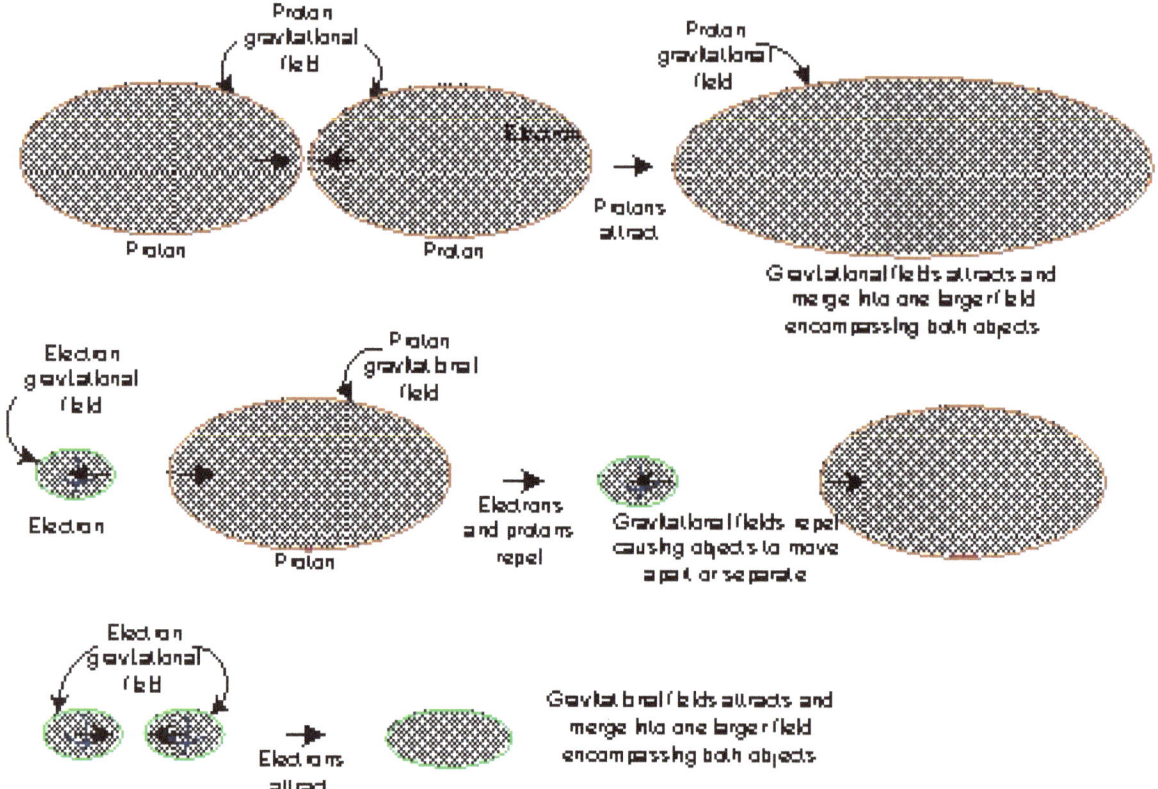

Figure 17.2. Protons and electrons have gravitational fields of opposite polarity. Two proton gravitational fields tend to merge into one, so two protons attract each other. Two electron gravitational fields also tend to merge, so electrons also attract each other, but proton and electron gravitational fields repel each other and thus the particles tend to separate from each other. The magnitude of the force between two particles (attractive or repulsive) depends on the size of the gravitational field, which in turn depends on the mass and on the amount of photon energy inside the particles. The mass does not change but the amount of photon energy can. The strength with which two gravitational fields attract or repel depends only on the photon energy inside the particles, which is therefore also called gravitational energy (see Article 181: Stellar Cores and deciphering gravity and Article 182: Einstein's dream realized: unified field theory of electrogravitation) [1, 2].

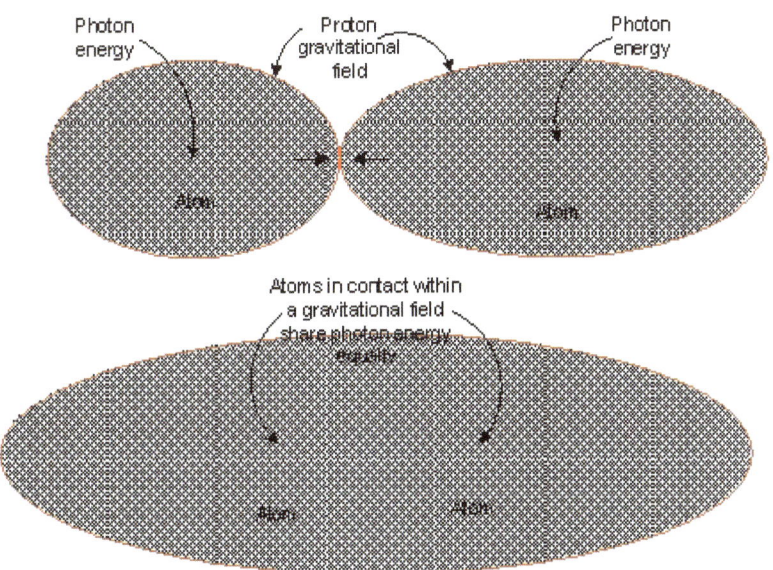

Figure 17.3. Two atoms with different amounts of gravitational energy will share the energy equally once they make contact. This contact seems to be (from observing Stellar Cores in the Sun's corona) that the two objects come close enough to each other so that the field of the lower energy object makes contact with the surface of the higher energy object.

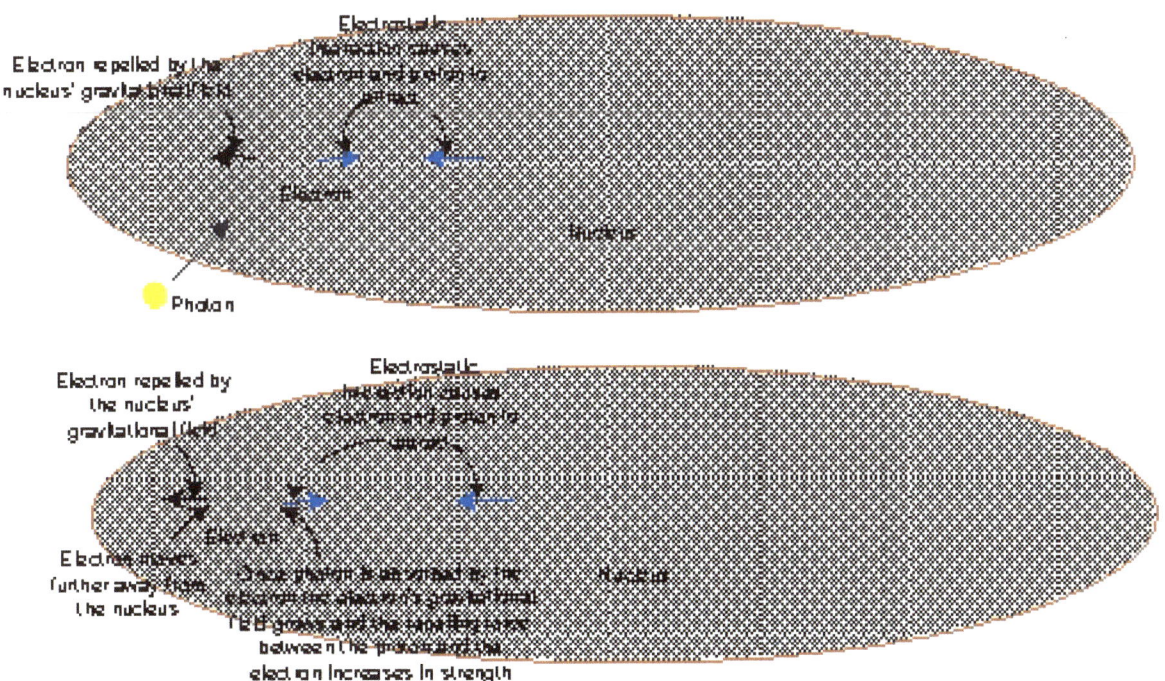

Figure 17.4. An electron is able to exist within the gravitational field of an atom even though the fields have opposite polarity, because the electrostatic interaction, between the nucleus (containing protons) and electrons, is attractive. If the electron absorbs photon energy then its gravitational field grows in

size and the gravitational repulsion increases. This results in the electron moving further away from the nucleus. If the electron gains enough photon (gravitational) energy, the strength of the repelling force between its field and the nucleus will be enough for it to leave the atom, which explains the process of ionization.

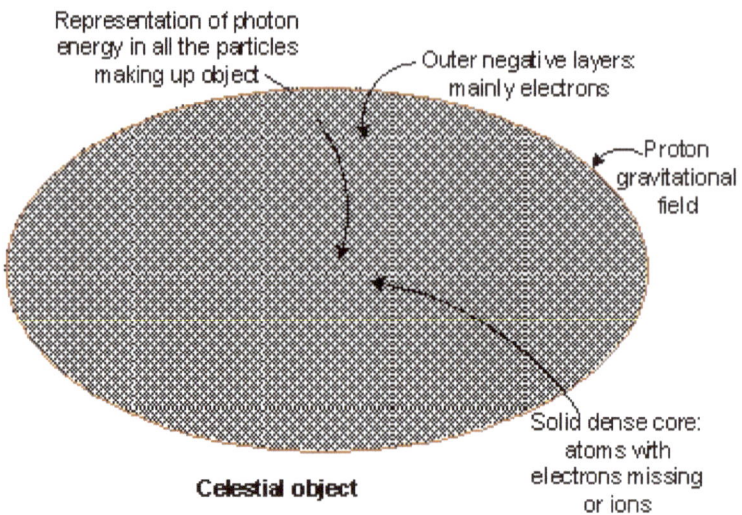

Figure 17.5. A celestial body will have a proton type of gravitational field because protons have larger gravitational fields, so this type of gravitational field will be predominant in the universe. Electrons' gravitational fields are repelled by a proton gravitational field, which is densest in the core of a celestial object, as this is the densest part of the object and where protons are the closest to each other. The repulsion between proton and electron gravitational fields causes electrons in the outermost part of atoms to be pushed away from their atoms and thus end up in the object's outermost layer. They will not leave the object because the electrostatic force keeps them bound to the positively charge inner core.

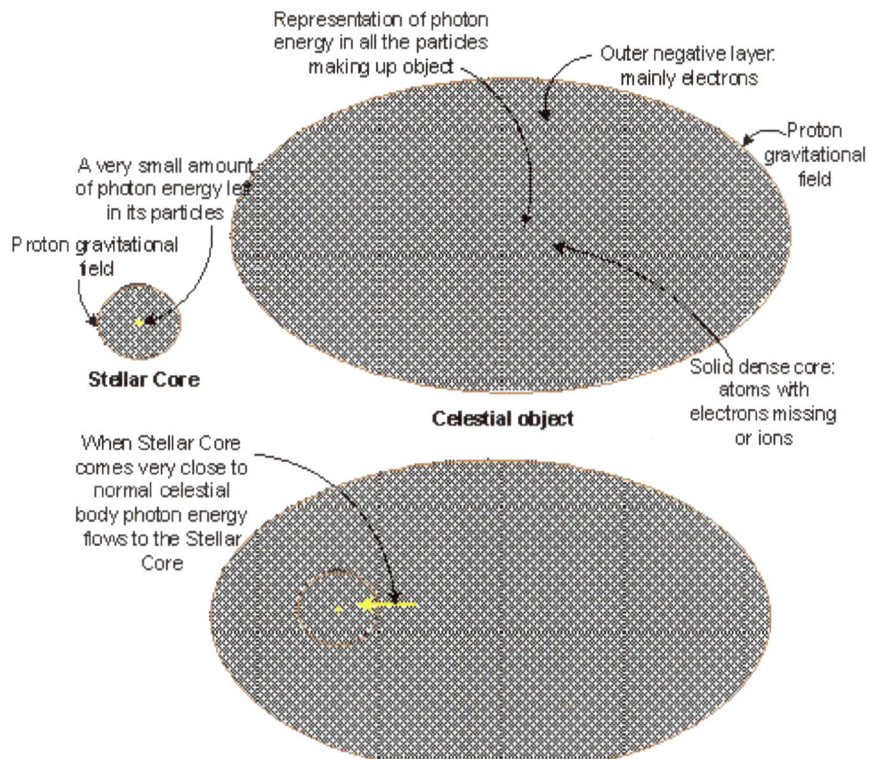

Figure 17.6. A Stellar Core has no negative outer layer, a very small amount of photon energy and a very small proton type gravitational field (see Article 184: Stellar Core evolution) [3]. When a Stellar Core makes contact with a normal celestial body by its gravitational field making contact with that object's atmosphere (the corona, if the object is the Sun), photon or gravitational energy from the object's particles (represented by the yellow ellipse) flows to the Stellar Core.

The interaction between a Stellar Core and another celestial object is weak because the Stellar Core's gravitational field does not extend very far beyond its outer edge.

The difference between the gravitational force, exerted by an object of mass M on an object of mass m, a distance r apart, and the strength of the gravitational interaction, between the two, can also be understood from the universal law equation:

$$F = \frac{GMm}{r^2}$$

where F is the magnitude of the force and G is the gravitational constant which gives a measure of the strength of the gravitational field. The force depends on both G (depends on the photon energy of the object or particle) and the mass M of the object. The mass M gives a measure of the capacity the object has to hold photon energy and is thus a measure of the size of the object, and G is a measure of how far beyond the edge of the object the gravitational field extends, the further it extends the stronger the gravitational interaction and the larger G will be.

If an object has very low gravitational energy than its gravitational field will not extend very far from its surface or it may not even reach its surface. This means that any matter beyond the point where the object's gravitational field extends is not attracted to the object and it will be lost. This is the likely reason why Stellar Cores shed their outer layers. This is also the reason why the earth's surface is breaking up. The earth's gravitational field is decreasing in size, due to the Stellar Cores drawing on its gravitational energy, and therefore the earth's surface is no longer being attracted as strongly as before and the strength of the gravitational interaction it is able to have is decreasing (the earth's gravitational constant G_{Earth} is decreasing) so the earth expands and its surface breaks open.

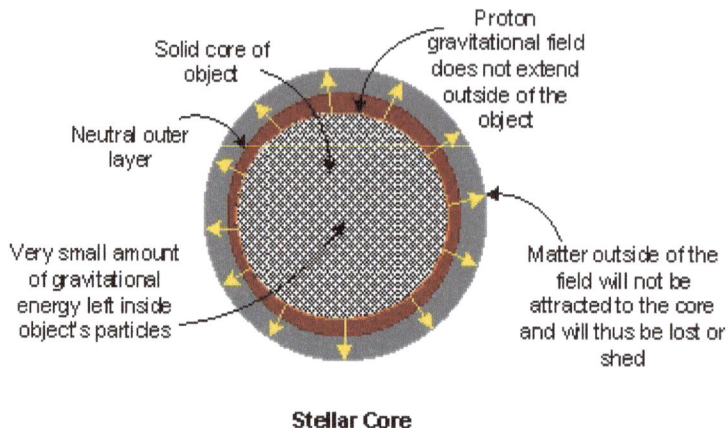

Stellar Core

Figure 17.7. Stellar Cores with very low photon energy inside them will likely have gravitational fields, which do not extend to the edge of the object, which will cause all the material beyond the edge of the gravitational field not to be attracted and is thus lost or shed. This is the reason why Stellar Cores shed their outer layers of material and have filled the Solar System with debris that just floats around.

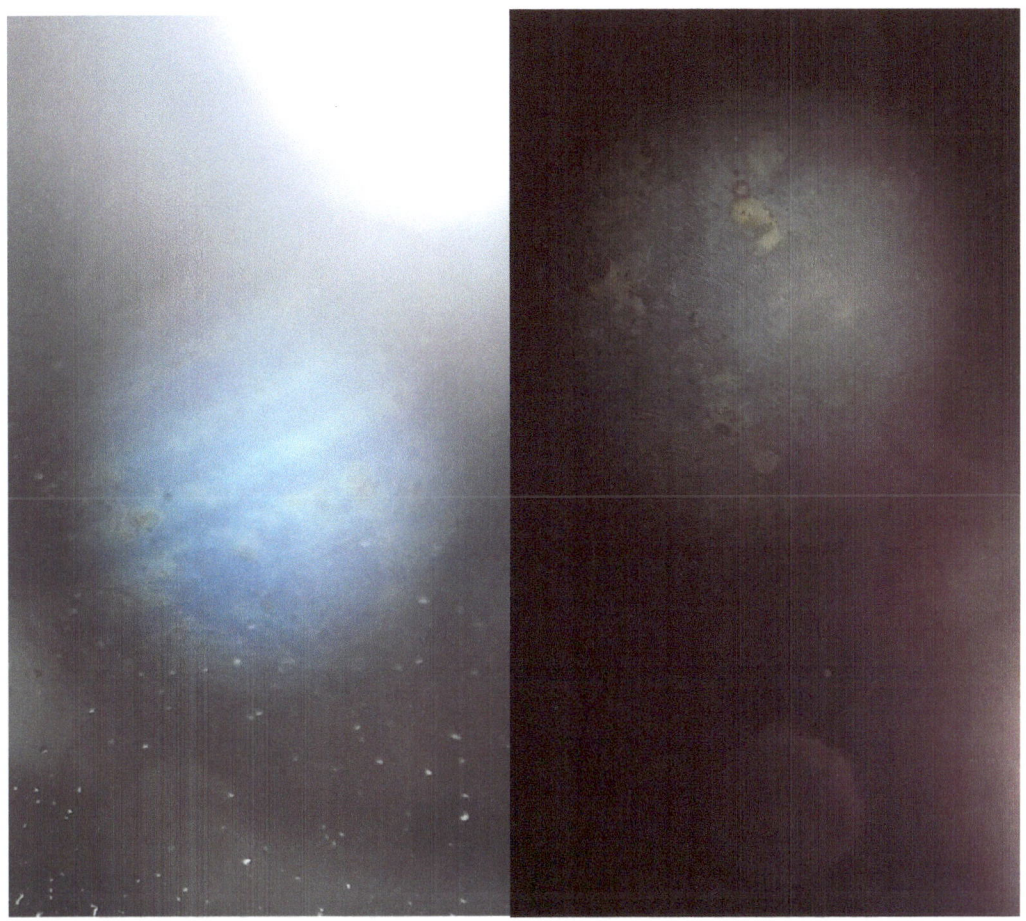

Figure 17.8. On the left: the Telescopic image of the Blue Stellar Core from May 12th, 2017 showing that the material making up its stripes is solid but clumpy as it is obviously being shed in small clumps. The debris resulting from the shedding of this material floats around the object. The debris is obviously not affected by any gravitational interaction (G = 0) indicating that it has zero photon or gravitational inside it. On the right: Photographic image of the same object from July 26th, 2017 showing that it had shed almost all the remaining material clinging to its surface.

Figure 17.4. Stellar Core in a Stereo COR2 image from November 30th, 2017. The object is approximately half the size of the Sun. Debris (black and white specks) can also be seen just floating around and thus matter with zero gravitational interaction strength.

In conclusion, celestial object interactions are mostly large scale versions of what occurs at the microscopic level with particles. There is gravity (proton gravitational fields) and antigravity (electron gravitational fields). Stellar Cores absorb a celestial object's photon or gravitational energy by coming very close to its surface so that the normal mechanism of photon energy sharing occurs. However, the flow of energy will most likely be much slower than with particles so that the Stellar Core has to come back over and over again to repeat the process. The gravitational interaction depends on the size of an object's gravitational field. The strength of the gravitational interaction depends on how far beyond the edge of an object its gravitational field extends. Gravity emerges from photons and is powered by photons, so gravity is truly the photon field.

References:

[1]   Albers, C. (2017). Article 181: Stellar Cores and deciphering gravity

[2]   Albers, C. (2018). Article 182: Einstein's dream realized: a unified field theory of electrogravitation.

[3]   Albers, C. (2017). Article 184: Stellar Core evolution.

Chapter 18

250:. Planet X causing Earth to move away from the Sun

The Sun has been drained of gravitational energy for over 150 years (see Article 146: Planet X System: time of arrival) [1] by the Planet X System of gravitational energy depleted, or dead, stars and planets, also called Stellar Cores. The name Stellar Core is appropriate because what is left of these objects is the solid dense core of the object and a small amount of material around that core. The rest of the material that was once a part of the living celestial objects, star or planet, would have been lost long ago. This occurs because energy depleted celestial objects lose their gravity. The gravity or the ability an object has to attract its own matter and other objects are dependent on their gravitational energy, which is generated through radioactive decay in the core (see Article 192: Neutron stars and fission as a star's internal energy source) [2]. Once there are no more nuclei to decay in the core, the object becomes energy depleted, which is analogous to its internal battery has run out of energy. Once the energy is depleted gravity sharply declines (see Article 249: The photon field: Gravity, antigravity and gravitational energy) [3]. In addition, gravitational energy depleted objects absorb energy from objects that are not energy depleted and this is the reason why the Sun and all the planets in the Solar System are being drained by the objects belonging to the Planet X System, which have invaded our Solar System (see Article 244: Planet X System: destroyer of star systems) [4].

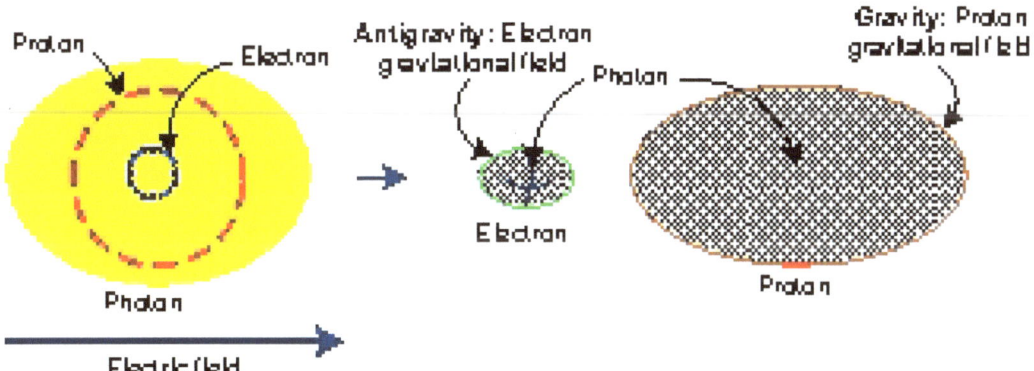

Figure 18.1. All matter and particles in the universe come from photons. The gravitational field of an object also comes from photons and in particular, the strength of the gravitational interaction is dependent on the amount of photon energy inside a particle or celestial object (a very large combination of particles) (see Article 249: Gravity, antigravity, and gravitational energy) [3]. The evidence that this occurs comes from particle anti particle creation, when for example a gamma ray splits into an electron positron pair, except that there is no anti mass, only mass positron is simply a particle with the same mass as the electron but with a positive charge, the two particles quickly recombine into a gamma because the strength of the gravity and antigravity interaction is the same, in this case (see Article 182: Einstein's dream realized: unified field theory of electrogravitation) [5]. The accepted gravitational force (force between planets and the Sun) is due to the gravitational field

generated by protons, the gravitational field generated by electrons has opposite polarity and is therefore called antigravity [3].

As the object being drained loses gravitational energy and the gravitational interaction strength, *G* (called the gravitational constant in Newton's gravitational law, but which is not a constant), drops (see Article 210: Stellar Core gravity: tidal and G is not constant) [6]. The gravitational constant is, therefore, constant for one particular celestial object (star or planet) but is not universally constant across the universe. The evidence that Stellar Cores interact through a much weaker gravitational constant comes from the 2007 Stellar Core which traversed the Sun in February of 2007.

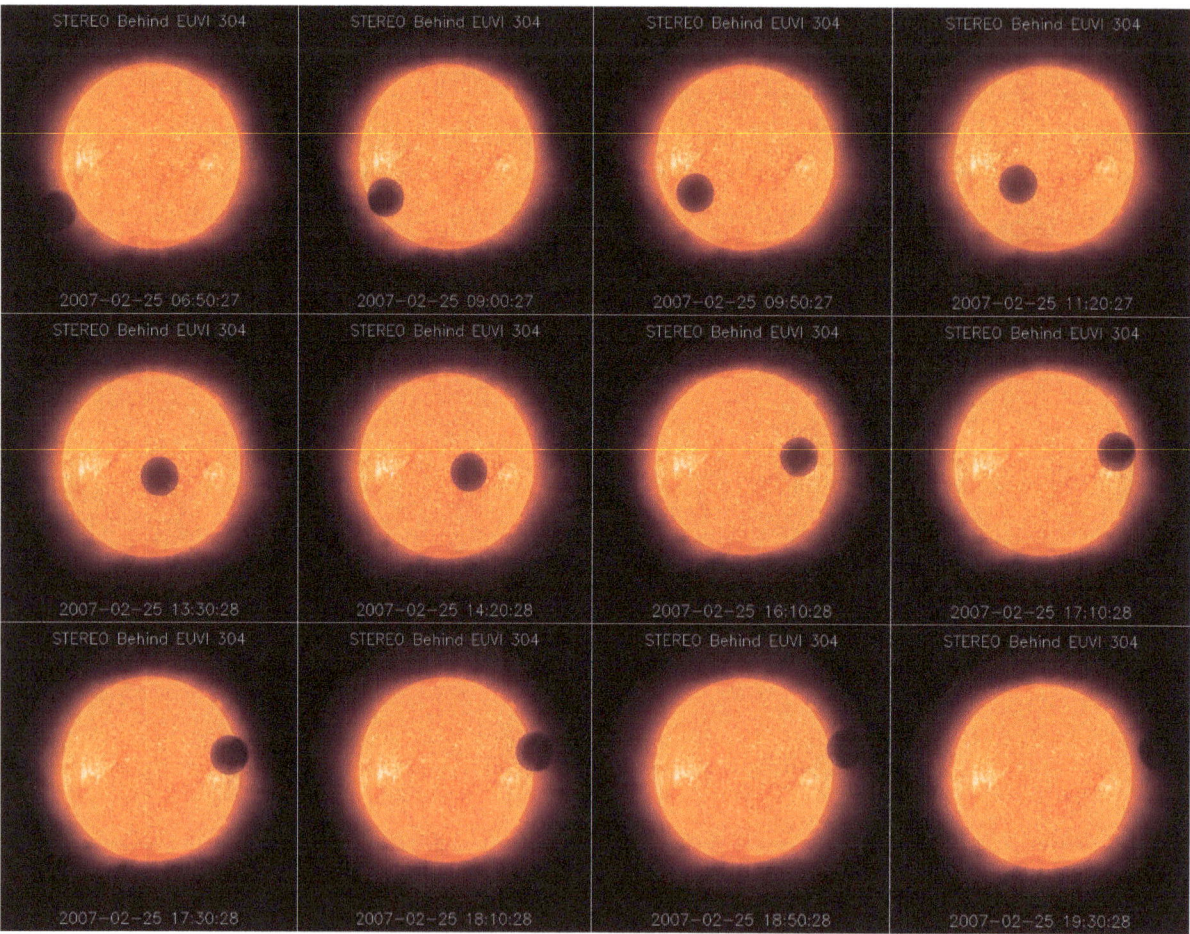

Figure 18.2. Stereo B EUVI 304 angstrom wavelength image from 2007: the 2007 Stellar Core traverses the Sun in the Sun's corona. The object was 2.2 times the size of Jupiter (see Article 143: Planet X traverses the Sun: irrefutable evidence) [7]. An object further from the Sun than Sun's inner corona would loop black in the images. This object looks and as it is entering and leaving the corona, the part of the object outside the corona looks black. The object takes 10 hours to traverse the Sun and is travelling at 39 km/s or 24 mi/s, or at a much lower speed than the Sun's escape velocity of 616 km/s (see Article 153: Planet X: Escape velocity and Gravity) [8].

This causes the celestial object to expand in size and for any satellite to move further away from it. So the fact that the Sun has been drained of gravitational energy would mean that all the planets in the Solar System would move away from the Sun. And since the Sun shows signs of being weakened at this time, as it is losing the corona and is much darker than it was a few years back at most wavelengths detected by the SDO satellite (see Article 195: Stellar Cores and the dying Sun) [9], it is very likely that most of the planets have moved further away from the Sun.

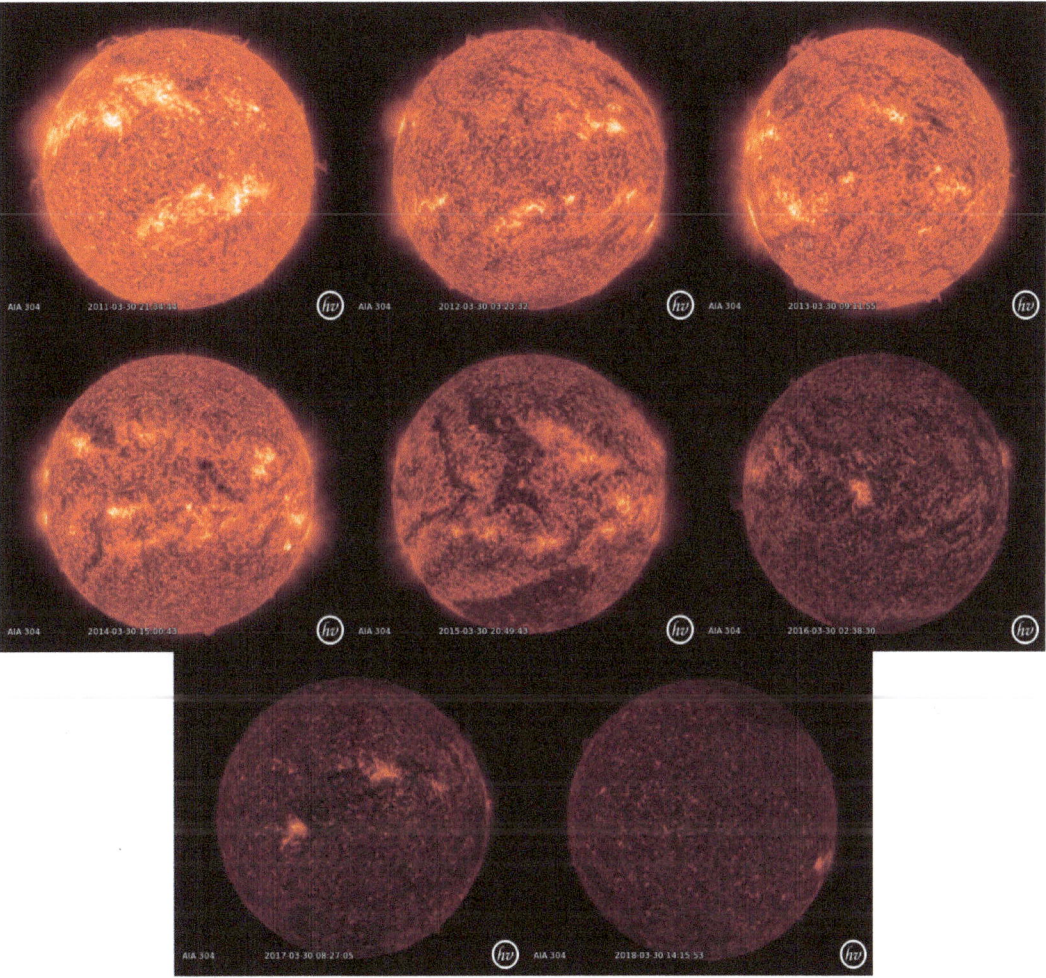

Figure 18.3. SDO images of the Sun in the 304 angstrom wavelength from 2011 to 2018. The Sun's has darkened dramatically showing that it has been drastically weakened.

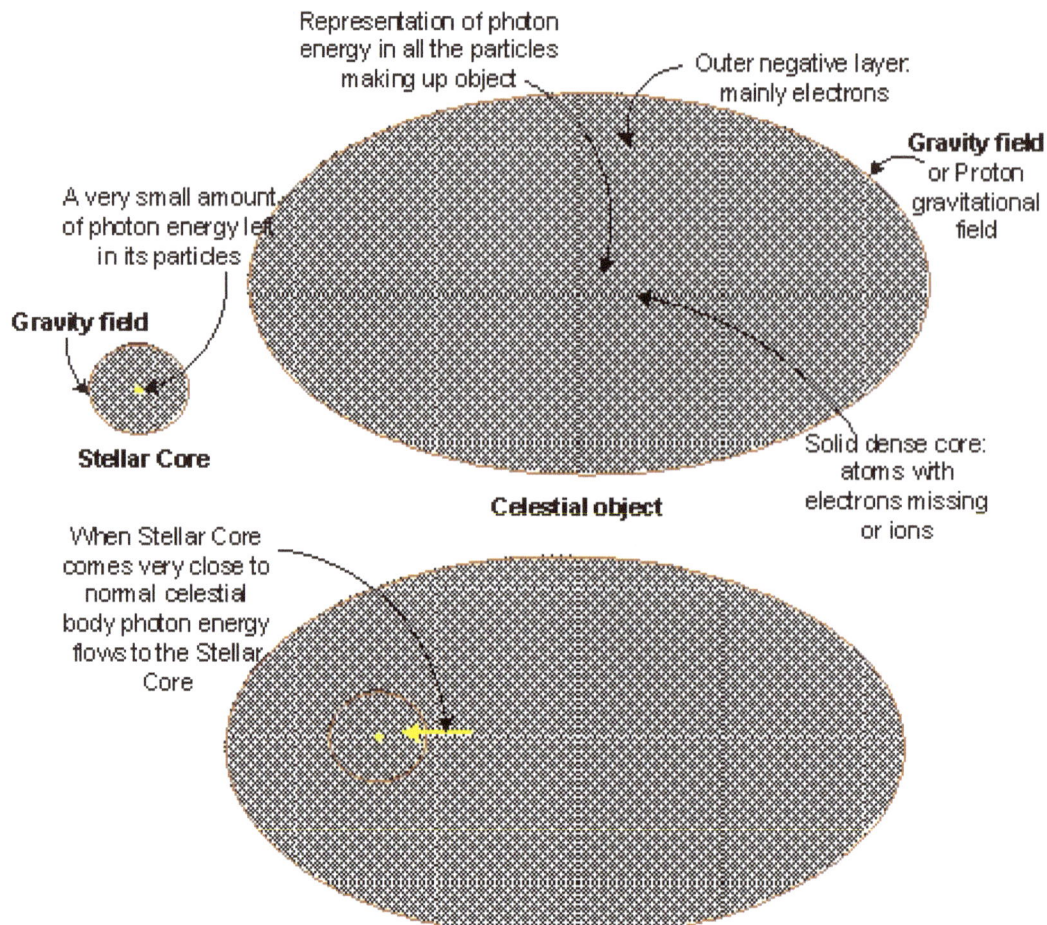

Figure 18.4. The mechanism behind Stellar Cores interacting very weakly through the gravitational interaction: Stellar Core gravitational fields are very weak (due to low internal energy) and thus do not extend very far from the edge of the object. Stellar Cores drain energy from the Sun by making contact with it with their gravitational field.

In order to see this, we only need to use Kepler's second and third laws, in regards to the motion of the planets in the Solar System. Kepler's second law basically states that a planet's angular momentum is constant or conserved, which translates into the equation:

$$L = r^2 v \quad \Rightarrow \quad v = \frac{L}{r^2} \tag{1}$$

where r is a planet's orbital radius, v is its orbital speed and L is its angular momentum per unit mass and we will assume that it remains constant as the Sun loses gravitational energy. Now, one of the ways we can write Kepler's 3rd law is:

$$\frac{GM}{r} = v^2 = \frac{L^2}{r^4} \quad \Rightarrow \quad r^3 = \frac{L^2}{GM} \tag{2}$$

This equation implies that if G decreases than r increases. If the Sun has lost 30% of its original gravitational energy, then we may expect to have: $G_{new} = 0.7G$. Thus

$$\frac{r_{new}^3}{r^3} = \left(\frac{L^2}{G_{new}M}\right) / \left(\frac{L^2}{GM}\right) = \frac{G}{G_{new}} = \frac{1}{0.7} \Rightarrow r_{new} = \frac{r}{\sqrt[3]{0.7}} = 1.13r \quad (3)$$

This would mean that the earth would now be 13% further from the Sun than is normal. But a planet's orbital period or the time it would take to complete an orbit would also increase so the length of the year would increase, which would cause the seasons, for example, Spring to arrive later. We can see this effect from another form of Kepler's third law:

$$T^2 = \frac{4\pi^2}{GM}r^3 = \frac{4\pi^2}{GM}\frac{L^3}{GM} \Rightarrow T = \frac{2\pi L}{GM} \quad (4)$$

where equation (2) was substituted into equation (4). Now, if again we consider that the Sun may have lost 30% of its energy so that $G_{new} = 0.7G$, we get:

$$\frac{T_{new}}{T} = \left(\frac{2\pi L}{G_{new}M}\right) / \left(\frac{2\pi L}{GM}\right) = 1.4 \quad (5)$$

which means that the year would be 40% longer.

The fact that the Sun is losing gravitational energy to the Stellar Cores indicates that the Sun's gravity will be affected and that this will result in a reduced strength of the gravitational interaction, which is given by G. In which case, the Earth is now further away from the Sun than before, which would cause the Sun to look smaller in the sky, from Earth. But this would be masked by the Global Sun Simulation system, which hides the presence of the Stellar Cores and the fact that the Sun periodically goes completely dark (see Article 226b: Sun simulating devices: the irrefutable evidence) [10]. In addition, the length of the year would be affected and the seasons would not come at the correct time.

The earth is not likely to cool down as a result of moving further away from the Sun as the earth is also being drained, which would cause a greater amount of heat to flow outwards from the core. This is because when earth's gravity decreases, as a result of the energy drain, or G decreases, the radioactive decay rate is likely to increase resulting in more energy or heat being released from the earth's core and flowing outwards through earth's many layers toward the surface and through the atmosphere (see Article 240: Planet X System effect on radioactive decay rate and heating of planets) [11].

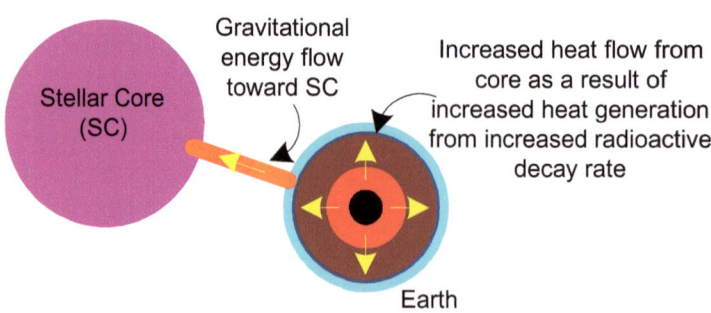

Figure 18.5. As energy flowing outwards from the core increases more energy flows through each layer of the earth, which causes each layer to experience an energy or heat, increase.

In conclusion, the fact that the earth is losing energy to the Stellar Cores is likely to increase both the orbital radius and the orbital period of the planets in orbit around the Sun. This will make the Sun look smaller in the sky, from earth, and the seasons to be delayed because the earth's orbit around the Sun will take longer to complete.

References:

[1]   Albers, C. (2018). Article 146: Planet X System: time of arrival.
[2]   Albers, C. (2018). Article 192: Neutron stars and fission as a star's internal energy source.
[3]   Albers, C. (2018). Article 249: The photon field: Gravity, antigravity and gravitational energy.
[4]   Albers, C. (2018). Article 244: Planet X System: destroyer of star systems.
[5]   Albers, C. (2018). Article 182: Einstein's dream realized: a unified field theory of electrogravitation.
[6]   Albers, C. (2018). Article 210: Stellar Core gravity: tidal and G is not constant.
[7]   Albers, C. (2018). Article 143: Planet X traverses the Sun: irrefutable evidence.
[8]   Albers, C. (2018). Article 153: Planet X: Escape velocity and Gravity.
[9]   Albers, C. (2017). Article 195: Stellar Cores and the dying Sun.
[10] Albers, C. (2018). Article 226b: Sun simulating devices: the irrefutable evidence.
[11] Albers, C. (2017). Article 240: Planet X System effect on radioactive decay rate and heating of planets.

Chapter 19

251: Active Object in inner Solar System tracked by the Stereo A Spacecraft Hi2 detectors

On a Stereo A Hi2 SREM image from May 26th, 2018, the planet Jupiter is observed approaching the occulter, which is always present in the Stereo A Hi2 images. The object which seemed to be a toroid space station caught by Scott C'one on May 14th, 2018 can also be seen below the occulter. Then as Jupiter reaches the occulter, a large particle ejection is seen coming from about Jupiter's position behind the occulter. But the ejection cannot be coming from Jupiter as Jupiter is not known to have the capability to create any such ejection and if it did it would be observed to have this type of ejection at other times and in different positions other than only behind the occulter.

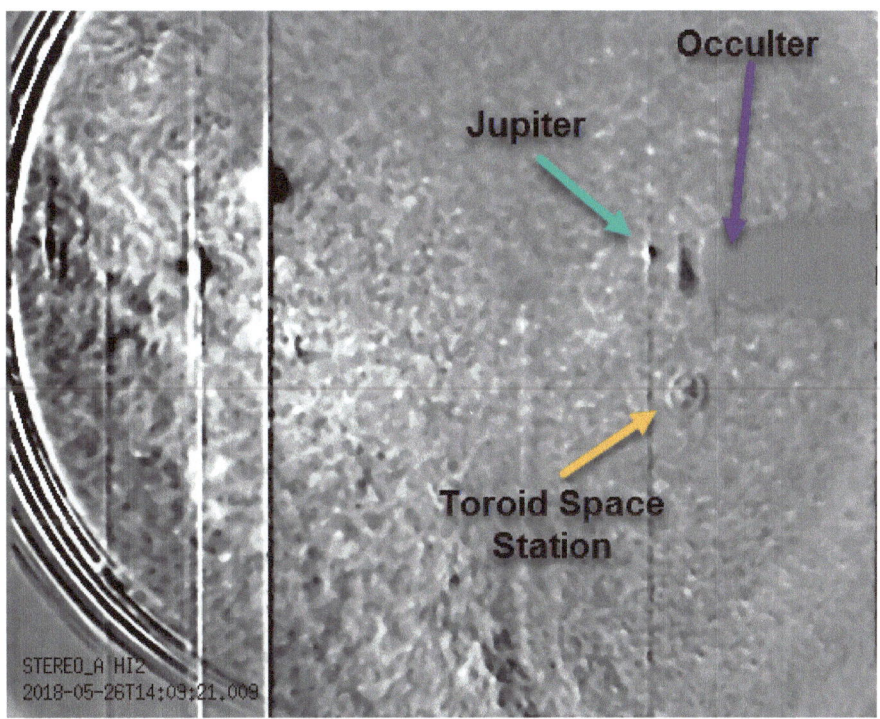

Figure 19.1. Stereo A Hi2 image from May 26th, 2018 showing the Toroid Space Station caught by Scott on May 14th, 2018 and the planet Jupiter approaching the occulter.

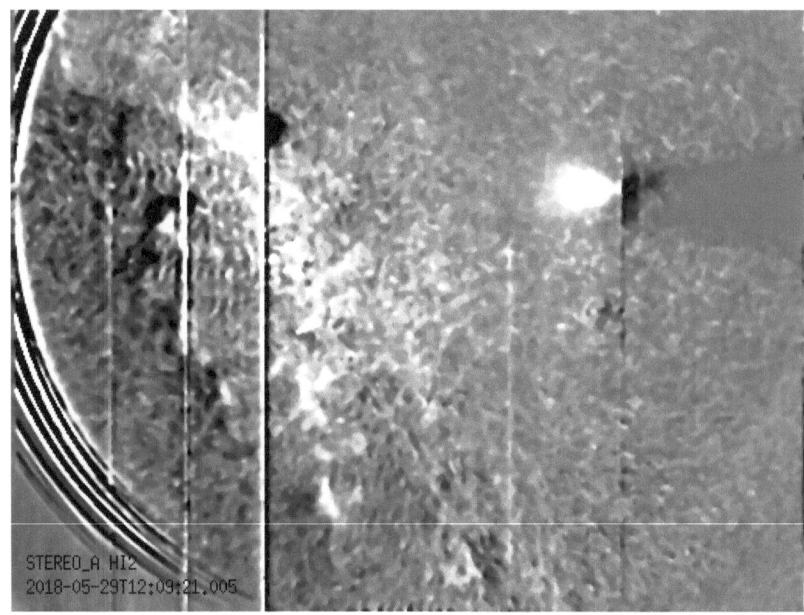

Figure 19.2. A large high energy particle ejection is seen issuing from Jupiter's position behind the occulter on May 29th, 2018.

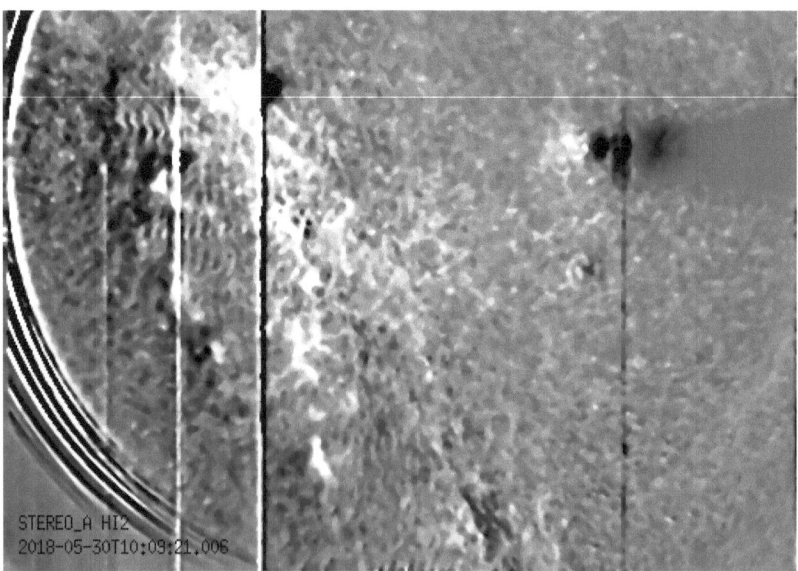

Figure 19.3. The discoloration on the occulter and behind Jupiter's position shows that there is an additional object present and that the occulter must be digitally added to the image so that particle emissions can still be seen right through the occulter.

The dark discoloration seen on the occulter is an indication that the occulter is digitally added to the image and that there is an object behind it emitting high energy particles as this is what the SREM detector is designed to detect, i.e. high energy protons, electrons and heavy ions, most of which will be helium nuclei. This means that there was an object behind the occulter before Jupiter reached the same position from the Stereo A spacecraft's point of view. This is an indication that the occulter is used to occult or hide an object which is tracked by the Stereo Spacecraft. The spacecraft would then be designed to keep the object in the position where the occulter is then digitally superimposed onto the image. It is also possible that some or all the planets are also superimposed onto the images. This would allow the operators to aim the spacecraft in different directions without alerting the public to that fact.

In the equivalent view, or the Hi2 wide angled view, but visible light, images, shown below, bright light emission can be seen to be coming from behind the occulter when Jupiter has not yet reached the occulter, which again indicates the presence of an object that must be responsible for the observed light emission.

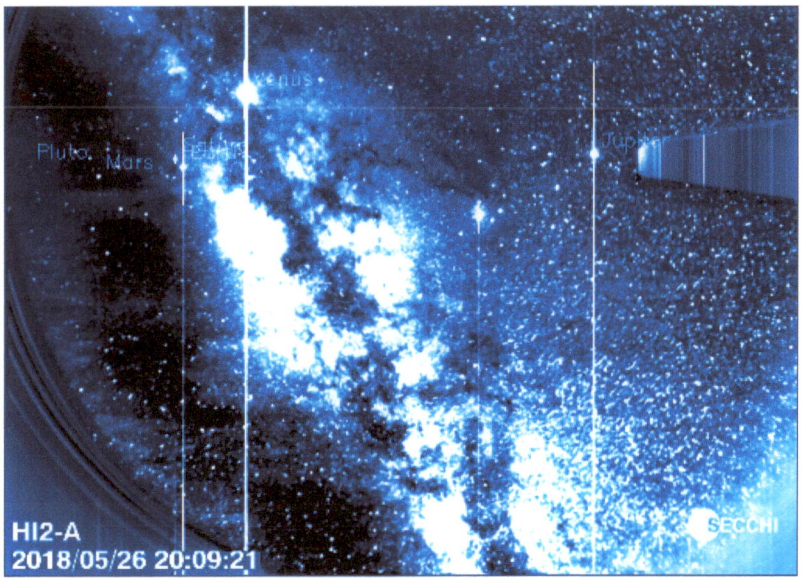

Figure 19.4. Bright light emission is observed coming from behind the occulter on May 26th, 2018, before Jupiter gets to the position, on the image, where the occulter is digitally superimposed.

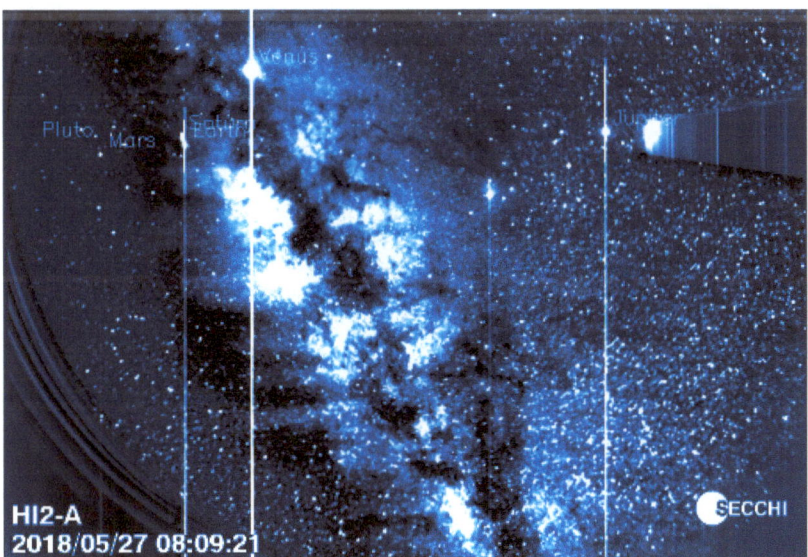

Figure 19.5. The bright light emission behind the occulter grows in size whilst Jupiter still has not reached the occulter's edge.

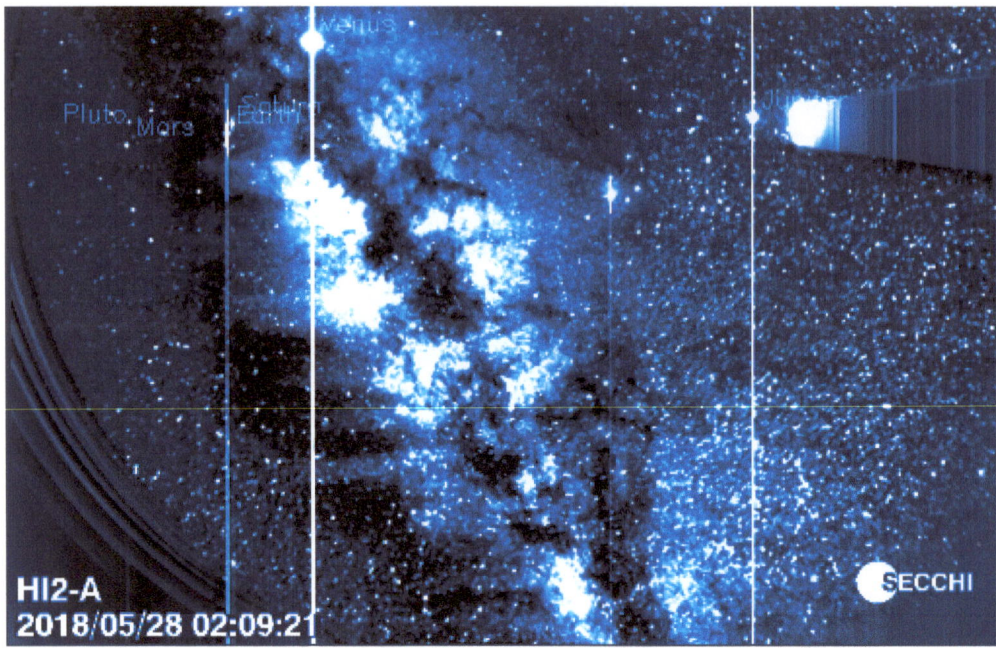

Figure 19.6. The region from which bright light is being emitted grows even larger in size. This again indicates that there has to be an object present at this position that is the source of the light emission.

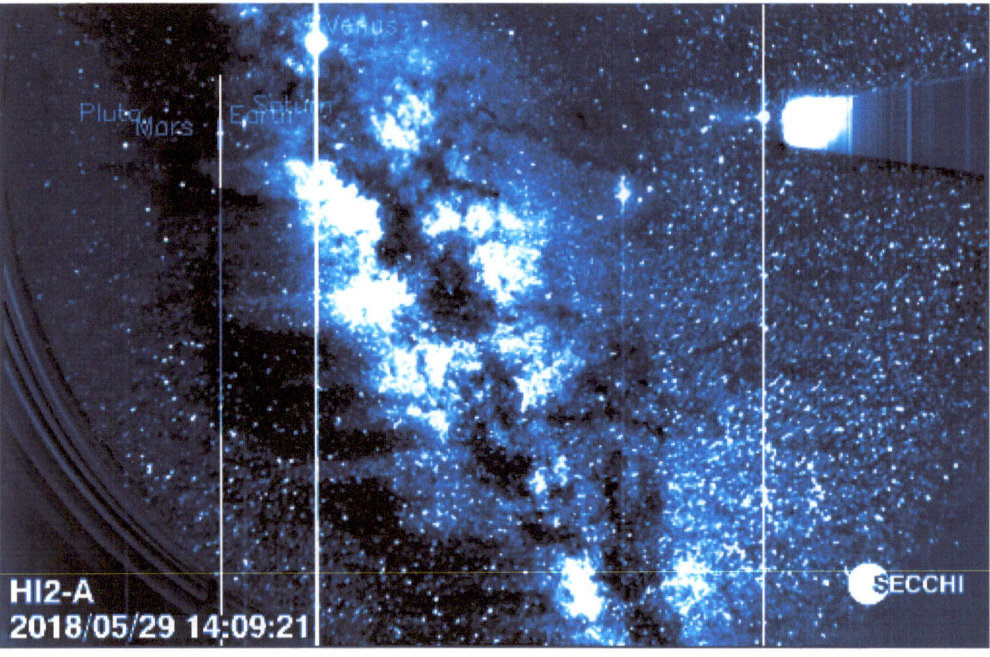

Figure 19.7. Having grown even larger, the light emission region behind the occulter is then seen to eject material, which emits light, indicating the presence of an active celestial object in a position behind the occulter.

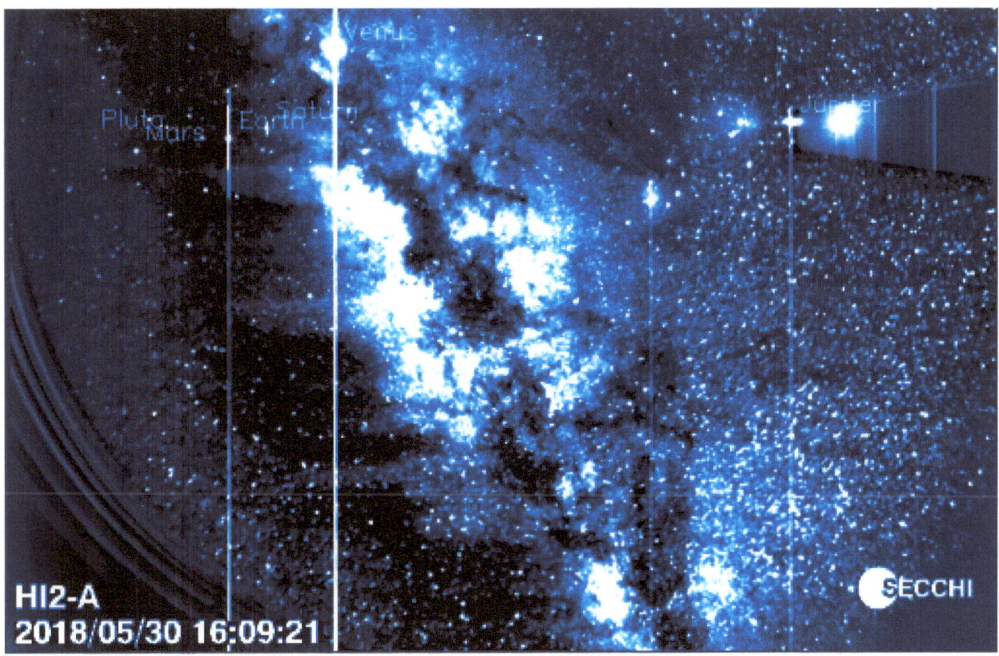

Figure 19.8. The bright light emission region diminishes in size whilst the ejection of light emitting plasma continues. There are 3 separate regions of increased brightness: Jupiter, a region in front of Jupiter and the occulter with 3 bright objects within it, and the region behind Jupiter, and behind the occulter.

Jupiter cannot, therefore, be the source of the ejection, an object seems to be present behind the occulter, which is capable of such an ejection, and the fact that it is situated behind an occulter, which is digitally added to the image, indicates that the object is purposely kept at the occulter's position. This, in turn, indicates that the Stereo A Hi2 detector is designed to keep it there and is thus tracking it.

In conclusion, the Stereo A spacecraft Hi2 detector seems to be designed to track an active celestial object, with the ability to eject large amounts of high energy particles. An occulter seems to be digitally added to the images in order to the occult, or hide, the object. The fact that the spacecraft is able to track the object from Earth's orbit whilst detecting what is to the left or to the right of the Sun suggests that the object is in the inner Solar System. This seems to, therefore, be a case of hiding something in plain sight.

Chapter 20

252: Planet X, cosmic rays, and human health

Cosmic rays in the earth's stratosphere, as reported by Spaceweather.com, on June 10th 2018 [1] have risen by 13%, between March 2015 and May 2017. This is a huge increase. Planet X Objects or Stellar Cores, also sometimes referred to as Brown Dwarf stars, in the Solar System, and in the Sun's corona, appear to be the main reason for this increase as will be detailed in this article.

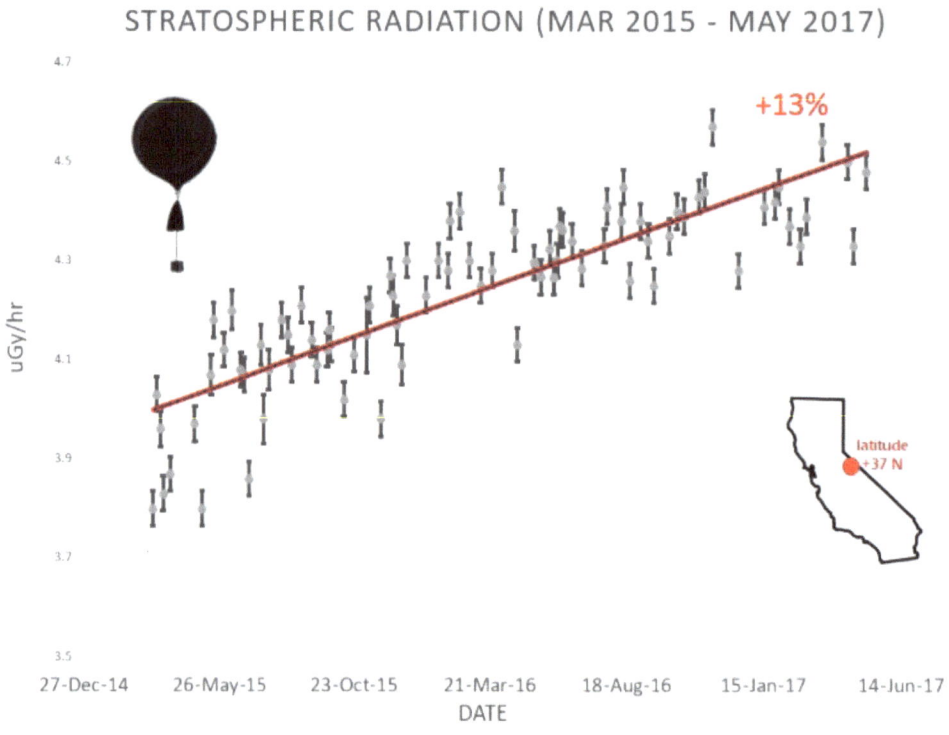

Figure 20.1. Cosmic ray data for the earth's stratosphere (60 000 ft) provided by spaceweather.com showing a 13% increase since March 2015 [1].

Cosmic rays are any particle, with the energy of 100 MeV or more. A cosmic ray particle travels at a speed which is greater than 40% of the speed of light. In other words, these are relativistic particles. These particles go right through the earth's magnetosphere. Cosmic rays are made up of mostly ionized hydrogen or protons (89%), some ionized helium (10%) and about 1% other nuclei such as iron, silicon, oxygen, carbon, and magnesium. The Sun is also a source of cosmic rays but when these come from the Sun they are called Solar Energetic Particles (SEPs).

When a cosmic ray, such as a high energy proton, impacts the atmosphere, it collides with an atmospheric particle, and it causes what is called an air shower. An air shower is the production of a cascade of secondary particles and gamma radiation. Some of the particles produced are the pion, symbolized by π (pi), which has a very short half-life, so it quickly decays to gamma rays and muons.

The gamma rays give rise to particle antiparticle creation, which results in the production of electrons and positrons (anti-electron) and also protons and antiprotons. An air shower, or cascade of particles, is shown in figure 2 below.

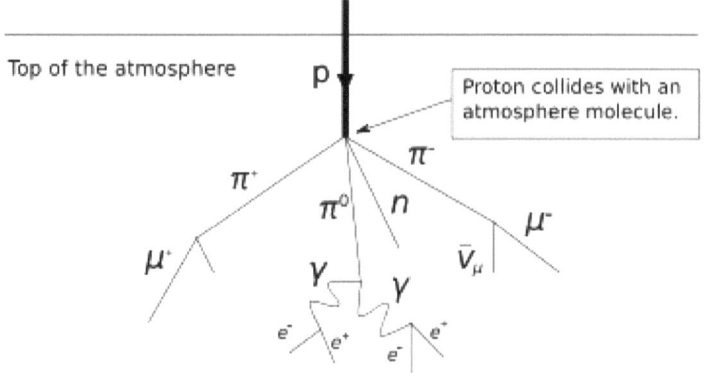

Figure 20.2. Illustration of an air shower, or cosmic ray particle cascade. P is a proton, π (pi) is a pion, μ (mu) is a muon, ν (nu) is a neutrino, n is a neutron, e⁻ (e minus) is an electron, e⁺ (e plus) is a positron (anti-electron), and γ (gamma) is a gamma ray.

Muons are stable enough to reach the earth's surface and cause damage to living organisms. They result in mutations and are therefore cancer causing. Cosmic rays deplete the ozone layer, and so increase the UVB (ultraviolet B or medium energy ultraviolet) radiation, emitted by the Sun, reaching the Earth's surface. UVB radiation is also cancer causing. The UVA radiation that our atmosphere allows through is not as damaging, but over exposure prematurely ages the skin and suppresses immunity, which may then lead to susceptibility to bacterial and viral infections. However, with the presence in the earth's atmosphere of a source of UVC radiation, this has raised the amount of UVC radiation reaching the earth's surface, to a level, which is now higher than would be expected had the earth completely lost its ozone layer, cosmic rays cannot, any longer, be considered to be the primary culprit for damage done by Ultraviolet radiation (see Article 204: Harmful UVC radiation reaching earth's surface indicates source within atmosphere) [2].

The usual source of cosmic rays is from outside the solar system. The Sun's solar wind produces a bubble, around the solar system, which deflects most of these particles. However, when the Sun weakens, as it has been doing, for many years now, due to the effect of Stellar Cores on it (see Article 109: Planet X and the weakening Sun) [3], the cosmic ray flux, impacting earth, also increases. Since the Sun is now weaker than ever, the amount of cosmic rays, from outside the Solar System, impacting earth has increased as the data shown in figure 1 indicates.

It has been accepted by solar physicists that the presence of large numbers of coronal holes, affect the number of SEPs, or solar cosmic rays, as any solar flare happening in an area adjacent to a coronal hole is likely to greatly accelerate particles coming through the coronal hole area [4]. What they have not however explained is that the presence of the Stellar Cores, in the Sun's corona, is what leads to both of these phenomena. The Stellar Cores draw coronal plasma from the Sun, thus creating coronal holes as well as causing the Sun's corona to continue to diminish in size and density. In addition, by drawing material from the Sun, they actually provoke the Sun into having CME events, in which a large number of particles, mainly positively charged ions are ejected from the Sun's corona. Thus, it is not surprising that the number of cosmic rays impacting earth will continue to increase. Stellar Cores affect the Sun in this way because first of all they drain the Sun of energy, which allows them to eventually emit light and increase their gravitational influence (see Article 250: Planet X causing Earth to move away from the Sun) [5], whilst the Sun's energy and gravitational influence decreases. With its gravitational influence decreased the Sun is less able to hold on to its material, and it will thus become easier for particles, to be accelerated during CME events, which leads to an increase in cosmic rays. The Sun's magnetic field would also be expected to decline, as the Sun weakens. And, the evidence that the Sun's magnetic field strength has been dropping comes from the observed decreasing magnetic field associated with sunspots.

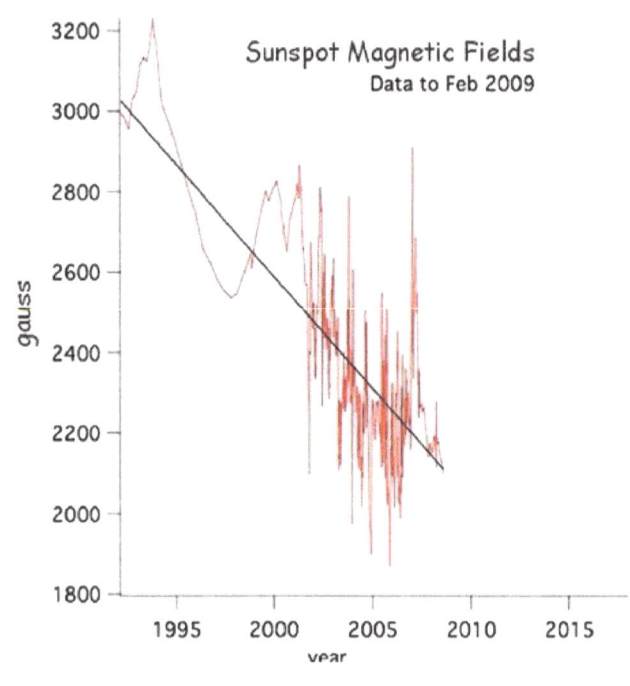

Figure 20.3. The Sun's magnetic field strength is dropping independent of the sunspot solar cycle [6].

In a study by Livingston and Penn [6], the Sun's magnetic field associated with sunspots, over the period between 1995 and 2009, is seen to drop consistently and independently of the solar cycle (see Article 118: Solar activity declining independent of the solar cycle: is the Sun dying?) [7]. The drop in the magnetic field strength was of about 50 gausses per year. Thus, the Sun's magnetic field has been consistently dropping since at least 1995, indicating that it has been weakening since then. This goes along with what can be observed from the fact that the Sun has grown much darker in SDO images over the years, in all ultraviolet wavelengths, but particularly in the 304 angstrom wavelength as shown below.

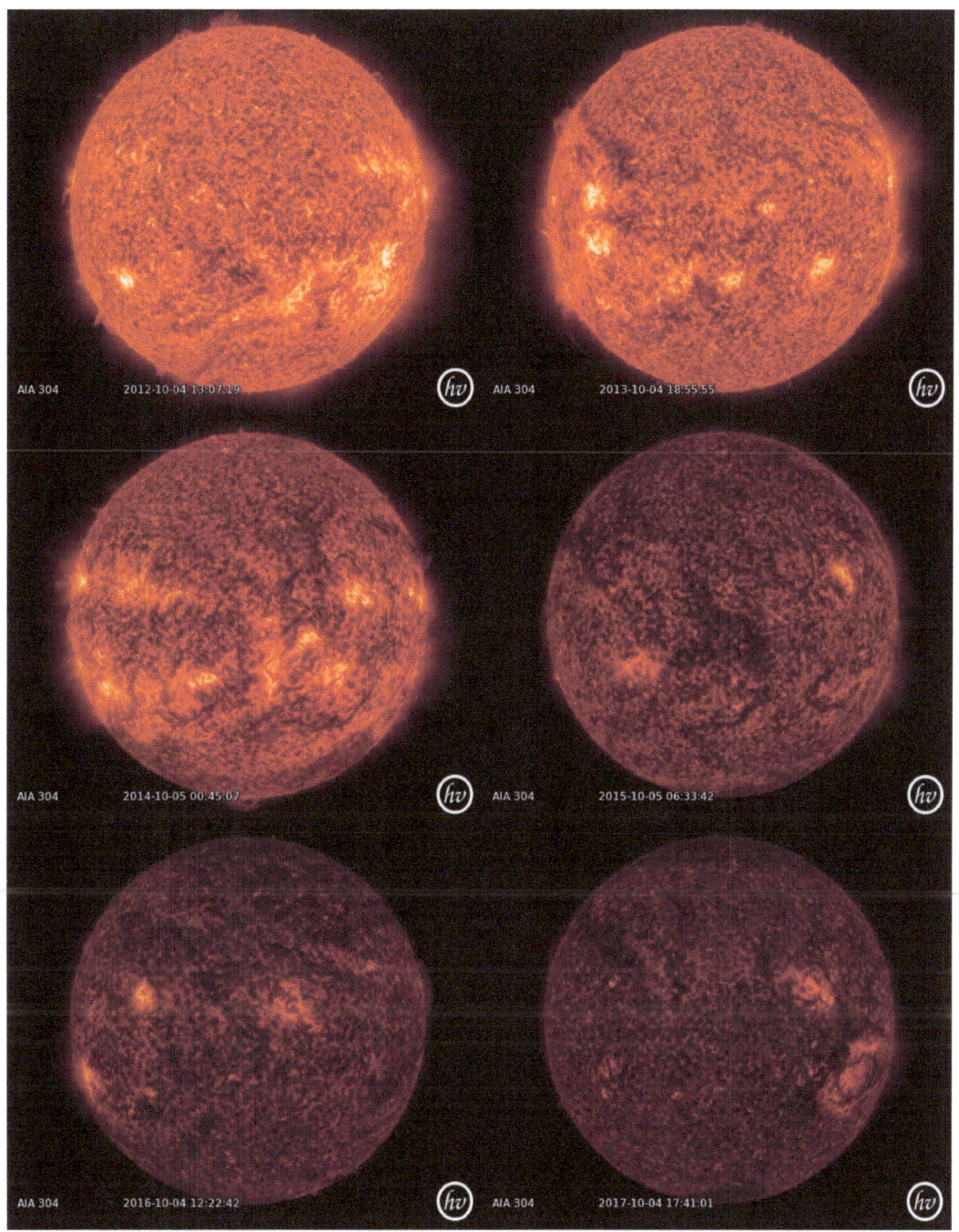

Figure 20.4. SDO images of the Sun in the 304 angstrom wavelength from 2012, 2013, 2014, 2015, 2016 and 2017. The Sun has clearly grown darker in this wavelength over the years. Comparison of the brightest spots indicates in the different images show that they can achieve the same lighter color and that therefore the increased darkness is not due to a change in the color assigned to different intensity. The 304 angstrom is emitted mostly from the upper chromosphere and this is, therefore, the Sun's layer that is seen in these images.

Thus, Planet X Objects or Stellar Cores are causing the cosmic ray flux increase in several ways:

1. The weaker Sun with its lower magnetic field has a decreased ability to deflect cosmic rays from outside the solar system, and thus more cosmic rays enter the solar system.
2. Stellar Cores connect through their weak gravitational influence, to the Sun, and draw material away from the Sun's upper layers. Once the material reaches a certain higher level, within the corona, it is prone to be repelled by the large amounts of electrons, in the Sun's corona, due to the gravitational repelling force between protons and electrons, and since the explosive ejection of positively charged ions, from the Sun's corona, is what causes a CME, and leads to solar flares, these objects actually provoke the Sun into having energetic events such as CMEs and solar flares (see Article 185: Stellar formation: Stars are formed from light) [8].
3. By drawing energy from the Sun (see Article 250: Planet X causing the Earth to move away from the Sun) [5], Stellar Cores weaken it, causing the Sun's gravitational influence to drop and thus making it easier for solar particles to experience higher accelerations and thus more easily reach a speed of 40% the speed of light, at which time they have become cosmic rays.

The increase in cosmic ray flux will off course have a dramatic effect on all living beings on earth, as cosmic rays can lead to DNA damage and thus cancer. Cosmic rays also appear to increase life-threatening arrhythmias in susceptible individuals [9]. But the first likely symptom of overexposure is tiredness and a weakened immune system. As the human body tries to find additional resources to cope with the DNA damage, occurring as a result of increased cosmic ray exposure, energy for other functions drops, so a person feels tired and the body may not have as much energy as before, to cope with bacterial and viral invaders. Since cosmic ray greatly increases with altitude, people who fly will be more prone to these effects. So what can a person do to help their body cope? Eating light meals rich in antioxidants, such as what is contained in blueberries, may help. Eating lightly is important because digestion of a heavy meal uses a lot of energy. Antioxidants help the body fight damage to cells and thus make more resources available to cope with cellular damage, created by cosmic rays. It is also important to keep hydrated, in other words, by drinking water, drinking other liquids may have the opposite effect. Water helps the body more easily get nutrients to the right places and waste out. For a person who is susceptible to heart arrhythmias or has a weak heart, it may be a good idea to take extra magnesium as a supplement, as magnesium relaxes the muscles and the heart is basically a muscle. In a more relaxed state, the heart is less likely to go into distress.

In conclusion, the dramatic cosmic ray flux increase, as reported by spaceweather.com, appears to be as a result of the presence of Planet X Objects or Stellar Cores, in the Inner Solar System, and in the Sun's corona. An increase in cosmic rays is expected to have a negative impact on living organisms, on earth, as this leads to DNA damage and cancer and heart arrhythmias in susceptible individuals.

References:

[1] http://spaceweather.com/ June 10th 2018

[2] Albers, C. (2017). Article 204: Harmful UVC radiation reaching earth's surface indicates source within the atmosphere.

[3] Albers, C. (2017). Article 109: Planet X and the weakening Sun.

[4] Kahler S.W. et al. (2013). Do Solar Coronal Holes Affect the Properties of Solar Energetic Particle Events? In: González Hernández I., Komm R., Pevtsov A., Leibacher J. (eds) Solar Origins of Space Weather and Space Climate. Springer, New York, NY.

[5] Albers, C. (2017). Article 250: Planet X causing Earth to move away from the Sun) [4.

[6] Matthew J. Penn and William Livingston, "Long-term Evolution of Sunspot Magnetic Fields." http://www. probeinternational.org/Livingston-penn-2010.pdf

[7] Albers, C. (2017). Article 118: Solar activity declining independent of the solar cycle: is the Sun dying?

[8] Albers, C. (2017). see Article 185: Stellar formation: Stars are formed from light.

[9] Stoupel, E. et al. (2008). The timing of life-threatening arrhythmias detected by implantable cardioverter-defibrillators in relation to changes in cosmophysical factors. Cardiol J. 15(5) pp. 437-440.

Chapter 21

253: Is the earth flat?

The earth, like all other astronomical bodies such the moon, other planets in our solar system, and the Sun, is spherical, in a three dimensional space or universe, but can be thought of as flat in a four dimensional space.

Figure 1 shows a ship, on the surface of the ocean, approaching an island, which is basically a very tall volcano so that a tall volcanic cone rises high, above the ocean surface. On a spherical earth, as the ship approaches the island, more and more of the volcanic cone can be seen from the ship's bridge, starting from the very top of the cone. When the ship is close to the island's beach, the whole cone and beach are visible. The view of the island from the ship's bridge is shown on the right hand side of figure 1. The island would also seem larger as the ship approaches it so we are going to assume that it is viewed through a telescope, so that its size is not important, only the percentage of the island which is visible above the surface of the ocean.

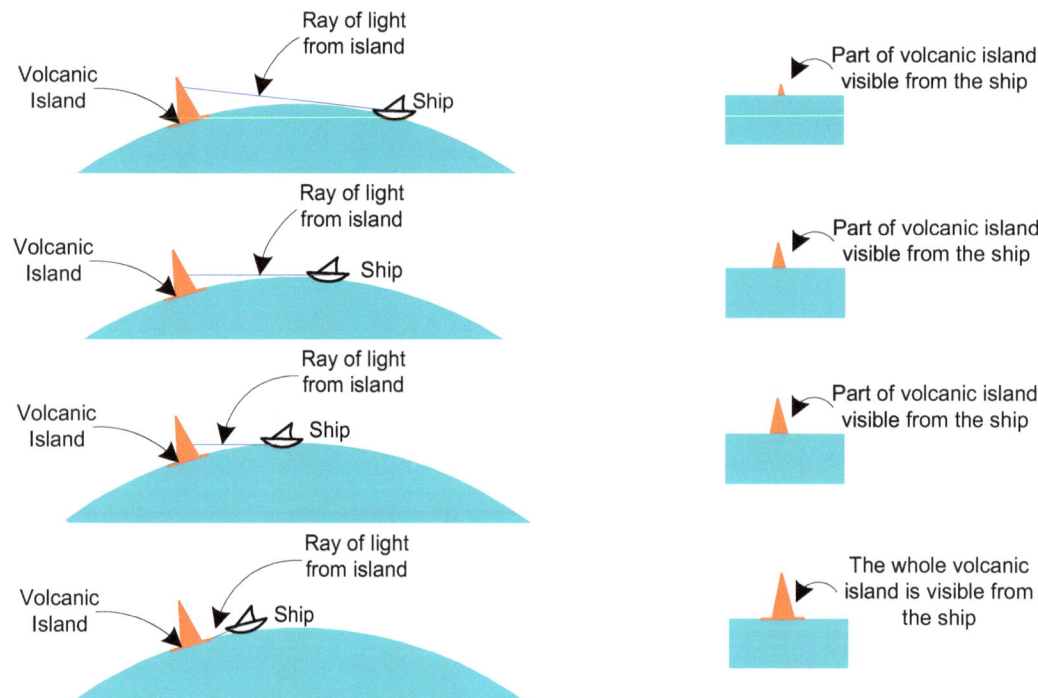

Figure 21.1. Illustration of how an island is seen from a ship on a spherical earth. On the left, the ship is at decreasing distances from the island, and on the right is the corresponding view of the island, from the ship.

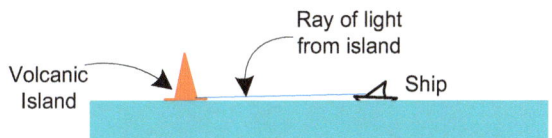 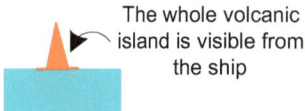

Figure 21.2. Illustration of the view from a ship, of a volcanic island, on a flat earth. The whole island can be viewed from the ship from very large distances.

However, if the earth was flat, no matter how far or close, the ship is from the island, the whole cone and beach would be visible through a telescope, from the ship's bridge, as long as ocean waves are not very high. This is illustrated in figure 2 above.

Since what is shown in figure 1 is what is actually observed from a ship, at sea, we have to conclude that the earth is spherical. However, the earth may be thought of as flat if viewed by a four dimensional being. Since we human beings can only perceive 3 dimensions, I will use a two dimensional world as an example. So, imagine a two dimensional world, or universe, which exists in 2 dimensions and contains two dimensional beings, and that these beings live on a planet, which has to be two dimensional, so we will use a circle to represent it. This two dimensional universe, in which these two dimensional beings live is shown in figure 3 below.

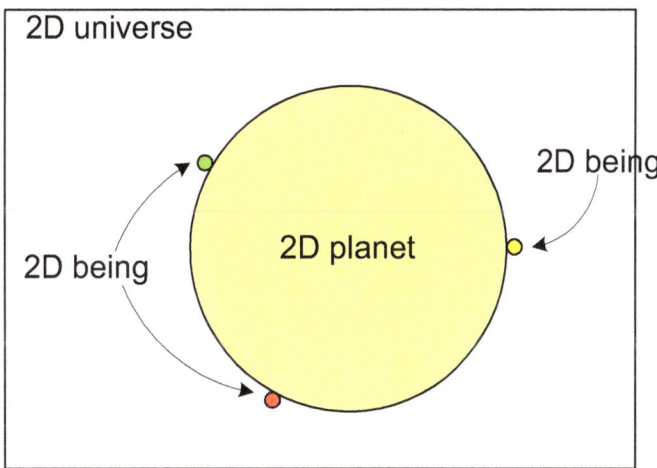

Figure 21.3. In a 2D (two dimensional) universe, a planet is a circle, and the beings, who live on it, are represented by colored circles.

The two dimensional beings have zero height, just like the planet, and so are not tall enough to see over the planet. So the beings cannot see each other if they are as far from each othe, as shown in figure 3. In a 2 dimensional world, light travels in a straight line but in the plane of the page.

Notice, therefore, from figure 3, that each of the beings cannot see each other and that if they want to get together they have to move around the circle, to the point, where the being they want to be with, is.

However, we are 3D (three dimensional) beings, living outside the 2D universe, in which the red, green and yellow beings live, so we can see all three of the beings at the same times, and are able to touch each one without moving around the circle. To us, 3D beings, this 2D planet is flat, but to the 2D beings living on the planet, it is curved or circular. In the same way, a 4D being would be able to see every point on our spherical 3D planet, at the same time, and thus our planet would, therefore, seem to be flat to them.

In conclusion, in the 3D (three dimensional) universe we live in, our planet is spherical, not flat. But if our planet could be viewed in 4D (four dimensions) it would seem to be flat.

The end for now!

www.ingramcontent.com/pod-product-compliance
Lightning Source LLC
Chambersburg PA
CBHW051912210526
45473CB00006B/1983